Domain-Specific Processors

Signal Processing and Communications

1. Digital Signal Processing for Multimedia Systems, *edited by Keshab K. Parhi and Takao Nishitani*
2. Multimedia Systems, Standards, and Networks, *edited by Atul Puri and Tsuhan Chen*
3. Embedded Multiprocessors: Scheduling and Synchronization, *Sundararajan Sriram and Shuvra S. Bhattacharyya*
4. Signal Processing for Intelligent Sensor Systems, *David C. Swanson*
5. Compressed Video over Networks, *edited by Ming-Ting Sun and Amy R. Reibman*
6. Modulated Coding for Intersymbol Interference Channels, *Xiang-Gen Xia*
7. Digital Speech Processing, Synthesis, and Recognition: Second Edition, Revised and Expanded, *Sadaoki Furui*
8. Modern Digital Halftoning, *Daniel L. Lau and Gonzalo R. Arce*
9. Blind Equalization and Identification, *Zhi Ding and Ye (Geoffrey) Li*
10. Video Coding for Wireless Communication Systems, *King N. Ngan, Chi W. Yap, and Keng T. Tan*
11. Adaptive Digital Filters: Second Edition, Revised and Expanded, *Maurice G. Bellanger*
12. Design of Digital Video Coding Systems, *Jie Chen, Ut-Va Koc, and K. J. Ray Liu*

13. Programmable Digital Signal Processors: Architecture, Programming, and Applications, *edited by Yu Hen Hu*
14. Pattern Recognition and Image Preprocessing: Second Edition, Revised and Expanded, *Sing-Tze Bow*
15. Signal Processing for Magnetic Resonance Imaging and Spectroscopy, *edited by Hong Yan*
16. Satellite Communication Engineering, *Michael O. Kolawole*
17. Speech Processing: A Dynamic and Optimization-Oriented Approach, *Li Deng and Douglas O'Shaughnessy*
18. Multidimensional Discrete Unitary Transforms: Representation, Partitioning, and Algorithms, *Artyom M. Grigoryan and Sos A. Agaian*
19. High-Resolution and Robust Signal Processing, *Yingbo Hua, Alex B. Gershman, and Qi Cheng*
20. Domain-Specific Embedded Multiprocessors: Systems, Architectures, Modeling, and Simulation, *Shuvra Bhattacharyya, Ed Deprettere, and Jürgen Teich*

Additional Volumes in Preparation

Biosignal and Biomedical Image Processing: MATLAB-Based Applications, *John L. Semmlow*

Watermarking Systems Engineering: Enabling Digital Assets Security and Other Applications, *Mauro Barni and Franco Bartolini*

Image Processing Technologies: Algorithms, Sensors, and Applications, *Kiyoharu Aizawa, Katsuhiko Sakaue, Yasuhito Suenaga*

Domain-Specific Processors

Systems, Architectures, Modeling, and Simulation

edited by

SHUVRA S. BHATTACHARYYA
University of Maryland
College Park, Maryland, U.S.A.

ED F. DEPRETTERE
Leiden University
Leiden, The Netherlands

JÜRGEN TEICH
University of Erlangen-Nuremberg
Nuremberg, Germany

MARCEL DEKKER, INC. NEW YORK • BASEL

Library of Congress Cataloging-in-Publication Data
A catalog record for this book is available from the Library of Congress.

ISBN: 0-8247-4711-9

This book is printed on acid-free paper.

Headquarters
Marcel Dekker, Inc., 270 Madison Avenue, New York, NY 10016, U.S.A.
tel: 212-696-9000; fax: 212-685-4540

Distribution and Customer Service
Marcel Dekker, Inc., Cimarron Road, Monticello, New York 12701, U.S.A.
tel: 800-228-1160; fax: 845-796-1772

Eastern Hemisphere Distribution
Marcel Dekker AG, Hutgasse 4, Postfach 812, CH-4001 Basel, Switzerland
tel: 41-61-260-6300; fax: 41-61-260-6333

World Wide Web
http://www.dekker.com

The publisher offers discounts on this book when ordered in bulk quantities. For more information, write to Special Sales/Professional Marketing at the headquarters address above.

Current printing (last digit):

10 9 8 7 6 5 4 3 2 1

PRINTED IN THE UNITED STATES OF AMERICA

Series Introduction

Over the past 50 years, digital signal processing has evolved as a major engineering discipline. The fields of signal processing have grown from the origin of fast Fourier transform and digital filter design to statistical spectral analysis and array processing, image, audio, and multimedia processing, and shaped developments in high-performance VLSI signal processor design. Indeed, there are few fields that enjoy so many applications—signal processing is everywhere in our lives.

When one uses a cellular phone, the voice is compressed, coded, and modulated using signal processing techniques. As a cruise missile winds along hillsides searching for the target, the signal processor is busy processing the images taken along the way. When we are watching a movie in HDTV, millions of audio and video data are being sent to our homes and received with unbelievable fidelity. When scientists compare DNA samples, fast pattern recognition techniques are being used. On and on, one can see the impact of signal processing in almost every engineering and scientific discipline.

Because of the immense importance of signal processing and the fast-growing demands of business and industry, this series on signal processing serves to report up-to-date developments and advances in the field. The topics of interest include but are not limited to the following:

- Signal theory and analysis
- Statistical signal processing
- Speech and audio processing

- Image and video processing
- Multimedia signal processing and technology
- Signal processing for communications
- Signal processing architectures and VLSI design

We hope this series will provide the interested audience with high-quality, state-of-the-art signal processing literature through research monographs, edited books, and rigorously written textbooks by experts in their fields.

Preface

Due to the rapidly increasing complexity and heterogeneity of embedded systems, a single expert can no longer be master of all trades. The era in which an individual could take care of all aspects (functional as well as nonfunctional) of specification, modeling, performance/cost analysis, exploration, and verification in embedded systems and software design will be over soon. Future embedded systems will have to rely on concurrency and parallelism to satisfy performance and cost constraints that go with the complexity of applications and architectures. Thus, an expert is familiar with and feels comfortable at only a few adjacent levels of abstraction while the number of abstraction levels in between a specification and a system implementation is steadily increasing. But even at a single level of abstraction, experts will most likely have good skills only in either computation- or communication-related issues, as a result of which the notion of *separation of concerns will become crucial.* These observations have far reaching consequences. One of them is that new *design methodologies* must be devised in which the notions of levels of abstraction and separation of concerns have grown into fundamental concepts and methods.

An early view on abstraction levels is represented by the *Y-chart* introduced by Gajski and Kuhn [1]. This chart gives three model views to *behavioral*, *structural*, and *physical*—showing levels of abstraction across which refinements take place. A more recent view on levels of abstraction and the relation between *behavior* and *structure* on these levels is reflected in the *Abstraction Pyramid* and the *Y-chart approach* introduced by Kienhuis et al. [2]. The Y-chart approach obeys a separation of concerns principle by making algorithms, architecture, and mapping manifest to permit quantification of choices. When looking more closely at the Gajski Y-chart and the Kienhuis Y-chart approaches one sees that the approach is invariant to the levels of abstraction: on each level there is some architecture or component that is specified in terms of some model(s) of architecture, there are one or more applications that are specified in terms of some model(s) of computation, and there are mapping methods that take components of the application model to components of the architecture model. The Y-chart, on the other hand, clearly reveals that models and methods will be different on each level of abstraction: refinements take place when going down the abstraction levels, as a consequence of which the design space is narrowed down step by step until only a few (Pareto optimal) designs remain.

Another consequence of the larger number of abstraction levels and the increasing amount of concurrency is that the higher the level of abstraction, the larger the dimension and size of the design space. To keep such a relation manageable, it is necessary to introduce *parametrized architectures* or templates that can be instantiated to architectures. The types of the parameters depend on the level of abstraction. For example, on the highest level, a method of synchronization and the number of a particular type of computing element may be parameters. Often, the template itself is a version of a *platform*. A platform is application-domain-specific and has to be defined through a domain analysis. For example, platforms in the automotive application domain are quite different from platforms in the multimedia application domain. In the former, platforms must match codesign finite state machine models [3], while in the latter they will have to support dataflow network or process network models [4].

Roughly speaking, a platform consists of two parts: one that concerns processing elements, and one that encompasses a communication and storage infrastructure. This partitioning is compliant with the rule computation vs. communication separation of concerns rule [5]. The processing elements are taken from a library—often as intellectual property components—and the communication and storage infrastructure is obeying certain predefined construction and operation rules. Specifying a platform is still more of an art than a science issue.

How could a sound methodology be designed to overcome the many problems that let to the paradigm change in embedded systems and software design? There is currently no compelling answer to that question. An increasing number of research groups all over the globe are proposing and validating prototype methodologies, most of which are embedded in the SystemC framework. An increasing number of them are advocating a platform-based design approach that relies heavily on models and methods that can support two major decision steps: exploration on a particular level of abstraction to prune the design space, and decomposition and composition to move down and up the abstraction hierarchy. The view that currently seems to prevail is a three-layer approach: an application layer, an architecture layer, and a mapping layer. The application layer and the architecture layer bear models of application(s) and the architecture, respectively, that match in the sense that a mapping of the former onto the latter is transparent. The mapping layer transforms application models into architecture models. Present approaches differ in the way the mapping of applications onto an architecture is conceived. One approach is to refine the application model to match the architecture model in such a way that only a system model, i.e., an implementation of the application, is to be dealt with when it comes to performance/cost analysis or exploration.

Another approach is to adhere strictly to the separation of concerns principles, implying that application models and architecture models are strictly separated. In this case, the mapping layer consists of a number of transformations that convert *representations* of components of the application model to representations of components of the architecture model. For example, a process in an application modeled as a process network can be represented by a control data flow graph (symbolically at higher [6] levels of abstraction and executable at lower levels of abstraction) and transformed subsequently in the mapping layer to come closer to the architecture processing unit model that supports various execution, synchronization and storage flavors.

This book offers a dozen essential contributions on various levels of abstraction appearing in embedded systems and software design. They range from low-level application and architecture optimizations to high-level modeling and exploration concerns, as well as specializations in terms of applications, architectures, and mappings.

The first chapter, by Walters, Glossner, and Schulte, presents a rapid prototyping tool that generates structural VHDL specifications of FIR filters. The object-oriented design of the tool facilitates extensions to it that incorporate new optimizations and techniques. The authors apply their VHDL generation tool to investigate the use of truncated multipliers in FIR filter implementations, and demonstrate in this study significant area improvements with relatively small computational error.

The next chapter, by Lagadec, Pottier, and Villellas-Guillen, presents a tool for generating combinational FPGA circuits from symbolic specifications. The tool operates on an intermediate representation that is based on a graph of lookup tables, which is translated into a logic graph to be mapped onto hardware. A case study of a Reed-Solomon RAID coder–decoder generator is used to demonstrate the proposed techniques.

The third chapter, by Guevorkian, Liuha, Launiainen, and Lappalainen, develops several architectures for discrete wavelet transforms based on their flowgraph representation. Scalability of the architectures is demonstrated in trading off hardware complexity and performance. The architectures are also shown to exhibit efficient performance, regularity, modularity, and amenability to semisystolic array implementation.

Chapter 4, by Takala and Jarvinen, develops the concept of stride permutation for interleaved memory systems in embedded applications. The relevance of stride permutation to several DSP benchmarks is demonstrated, and a technique is developed for conflict-free stride permutation access under certain assumptions regarding the layout of data.

The next chapter, by Pimentel, Terpstra, Polstra, and Coffland, discusses techniques used for capturing intratask parallelism during simulation in the Sesame environment for design space exploration. These techniques are based on systematic refinement of processor models into parallel functional units, and a dataflow-based synchronization mechanism. A case study of QR decomposition is used to validate the results.

The next chapter, by Hannig and Teich, develops an approach for power modeling and energy estimation of piecewise regular processor arrays. The approach is based on exploiting the large reduction in power consumption for constant inputs to functional units. An efficient algorithm is also presented for computing energy-optimal space-time mappings.

Chapter 7, by Derrien, Quillou, Quinton, Risset, and Wagner, presents an interface synthesis tool for regular architectures. Safety of the synthesized designs is assured through joint hardware/software synthesis from a common specification. Efficiency of the generated interfaces is demonstrated through experiments with DLMS filter implementation on a Spyder FPGA board.

In Chapter 8, by Lohani and Bhattacharyya, a model developed for executing applications with time-varying performance requirements and nondeterministic execution times on architectures with reconfigurable computation and communication structures. Techniques are developed to reduce the complexity of various issues related to the model, and a heuristic framework is developed for efficiently guiding the process of runtime adaptation.

The next chapter, by Turjan, Kienhuis, and Deprettere, develops and compares alternative realizations of the extended linearization model, which is an approach for reordering tokens when interprocess data arrives out of order during the execution of a Kahn process network (KPN). The model involves augmenting Kahn processes with additional memory and a controller in a manner that preserves KPN semantics. The alternative realizations are compared along various dimensions including memory requirements and computational complexity.

The next chapter, by Radulescu and Goossens, compares networks-on-chip with off-chip networks and existing on-chip interconnects. Network-on-chip services are defined and a transaction model is developed to facilitate migration to the new communication architecture. Properties of connections under the proposed network-on-chip framework include transaction completion, transaction orderings, performance bounds, and flow control.

Chapter 11, by Stravers and Hoogerbugge, presents an architecture and programming model for single-chip multiprocessing with high power/performance efficiency. The architecture is based on homogeneous clusters of processors, called tiles, that form the boundary between static and dynamic resource allocation. Case studies of an MPEG2 decoder are used to demonstrate the proposed ideas.

The last chapter, by Wong, Vassiliadis, and Cotofana, reviews the evolution of embedded processor characteristics in relation to programmability and reconfigurability. A case for embedded architectures that integrate programmable processors and reconfigurable hardware is developed, and a specific approach is described for achieving this by means of microcode.

We would like to thank all the chapter authors for their outstanding contributions. We also thank at Marcel Dekker Inc., B. J. Clark for his encouragement to develop this book and Brian Black for his help with the production process. Thanks also to all the reviewers for their hard work in helping to ensure the quality of the book.

Shuvra S. Bhattacharyya
Ed Deprettere
Juergen Teich

REFERENCES

1. Gajski, D. (1987). *Silicon Compilers*. Addison-Wesley.
2. Kienhuis, B., Deprettere, E. F., van der Wolf, P., Vissers, K. (2002). A methodology to designing embedded systems: the y-chart approach. In: Deprettere, E. F., Teich, J., Vassiliadis, S., eds. *Embedded Processor Design Challenges, Lecture Notes in Computer Science*. Springer.
3. Balarin, F., et al. (1997). *Hardware-Software Co-Design of Embedded Systems: The Polis Approach*. Kluwer Academic Publishers.
4. Kienhuis, B., Deprettere, E. F., Vissers, K., van der Wolf, P. (July 1997). An approach for quantitative analysis of application-specific dataflow architectures. In: *Proceedings of the International Conference on Application Specific Systems, Architectures, and Processors*.
5. Keutzer, K., Malik, S., Newton, R., Rabaey, J., Sangiovanni-Vincentelli, A. (December 19, 2000). System-level design: orthogonalization of concerns and platform-based design. *IEEE Transactions on Computer-Aided Design of Integrated Circuits and Systems*.
6. Zivkovic, V., et al. (1999). Fast and accurate multiprocessor exploration with symbolic programs. *Proceedings of the Design, Automation and Test in Europe Conference*.

Contents

Contributors

Shuvra S. Bhattacharyya University of Maryland at College Park, College Park, Maryland

Joe E. Coffland Department of Computer Science, University of Amsterdam, Amsterdam, The Netherlands

Sorin Cotofana Computer Engineering Laboratory, Electrical Engineering Department, Delft University of Technology, Delft, The Netherlands

Ed F. Deprettere Leiden Embedded Research Center, Leiden University, Leiden, The Netherlands

Steven Derrien Irisa, Campus de Beaulieu, Rennes, France

John Glossner Sandbridge Technologies, White Plains, New York, U.S.A.

Kees Goossens Philips Research Laboratories, Eindhoven, The Netherlands

David Guevorkian Nokia Research Center, Tampere, Finland

Anne-Claire Guillou Irisa, Campus de Beaulieu, Rennes, France

Frank Hannig University of Paderborn, Paderborn, Germany

Jan Hoogerbugge Philips Research Laboratories, Eindhoven, The Netherlands

Tuomas Järvinen Tampere University of Technology, Tampere, Finland

Bart Kienhuis Leiden Embedded Research Center, Leiden University, Leiden, The Netherlands

Loïc Lagadec Université de Bretagne Occidentale, UFR Sciences, Brest, France

Ville Lappalainen Nokia Research Center, Tampere, Finland

Aki Launiainen Nokia Research Center, Tampere, Finland

Petri Liuha Nokia Research Center, Tampere, Finland

Sumit Lohani University of Maryland at College Park, College Park, Maryland

Andy D. Pimentel Department of Computer Science, University of Amsterdam, Amsterdam, The Netherlands

Bernard Pottier Université de Bretagne Occidentale, UFR Sciences, Brest, France

Simon Polstra Department of Computer Science, University of Amsterdam, Amsterdam, The Netherlands

Patrice Quinton Irisa, Campus de Beaulieu, Rennes, France

Andrei Rădulescu Philips Research Laboratories, Eindhoven, The Netherlands

Tanguy Risset LIP, ENS Lyon, France

Michael J. Schulte ECE Department, University of Wisconsin–Madison, Madison, Wisconsin, U.S.A.

Paul Stravers Philips Research Laboratories, Eindhoven, The Netherlands

Jarmo Takala Tampere University of Technology, Tampere, Finland

Jürgen Teich University of Paderborn, Paderborn, Germany

Frank P. Terpstra Department of Computer Science, University of Amsterdam, Amsterdam, The Netherlands

Alexandru Turjan Leiden Embedded Research Center, Leiden University, Leiden, The Netherlands

Stamatis Vassiliadis Computer Engineering Laboratory, Electrical Engineering Department, Delft University of Technology, Delft, The Netherlands

Oscar Villellas-Guillen Université de Bretagne Occidentale, UFR Sciences, Brest, France

Charles Wagner Irisa, Campus de Beaulieu, Rennes, France

E. George Walters III CSE Department, Lehigh University, Bethlehem, Pennsylvania, U.S.A.

Stephan Wong Computer Engineering Laboratory, Electrical Engineering Department, Delft University of Technology, Delft, The Netherlands

Domain-Specific Processors

1

Automatic VHDL Model Generation of Parameterized FIR Filters

E. George Walters III
CSE Department, Lehigh University, Bethlehem, Pennsylvania, U.S.A.

John Glossner
Sandbridge Technologies, White Plains, New York, U.S.A.

Michael J. Schulte
ECE Department, University of Wisconsin–Madison, Madison, Wisconsin, U.S.A.

I. INTRODUCTION

Designing hardware accelerators for embedded systems presents many tradeoffs that are difficult to quantify without bit-accurate simulation and area and delay estimates of competing alternatives. Structural level VHDL models can be used to evaluate and compare designs, but require significant effort to generate.

This chapter presents a tool that was developed to evaluate the tradeoffs involved in using truncated multipliers in FIR filter hardware accelerators. The tool is based on a package of Java classes that models the building blocks of computational systems, such as adders and multipliers. These classes generate VHDL descriptions, and are used by other classes in hierarchical fashion to generate VHDL descriptions of more complex systems. This chapter describes the generation of truncated FIR filters as an example.

Previous techniques for modeling and designing digital signal processing systems with VHDL were presented in references 1–5. The tool described in this chapter differs from those techniques by leveraging the benefits of object oriented programming (OOP). By subclassing existing objects, such as multipliers, the tool is easily extended to generate VHDL models that incorporate the latest optimizations and techniques.

Subsections A and B provide the background necessary for understanding the two's complement truncated multipliers used in the FIR filter architecture, which is described in Section II. Section III describes the tool for automatically generating VHDL models of those filters. Synthesis results of specific filter implementations are presented in Section IV, and concluding remarks and given in Section V.

A. Two's Complement Multipliers

Parallel tree multipliers form a matrix of partial product bits which are then added to produce a product. Consider an m-bit multiplicand A and an n-bit multiplier B. If A and B are integers in two's complement form, then

$$A = -a_{m-1}2^{m-1} + \sum_{i=0}^{m-2} a_i 2^i \qquad \text{and} \qquad B = -b_{n-1}2^{n-1} + \sum_{j=0}^{n-2} b_j 2^j \tag{1}$$

Multiplying A and B together yields the following expression

$$\begin{aligned} A \cdot B = {} & a_{m-1}b_{n-1}2^{m+n-2} + \sum_{i=0}^{m-2}\sum_{j=0}^{n-2} a_i b_j 2^{i+j} - \sum_{i=0}^{m-2} b_{n-1}a_i 2^{i+n-1} \\ & - \sum_{j=0}^{n-2} a_{m-1}b_j 2^{j+m-1} \end{aligned} \tag{2}$$

The first two terms in Eq. (2) are positive. The third term is either zero (if $b_{n-1} = 0$) or negative with a magnitude of $\sum_{i=0}^{m-2} a_i 2^{i+n-1}$ (if $b_{n\text{-}1} = 1$). Similarly, the fourth term is either zero or a negative number. To produce the product of $A \times B$, the first two terms are added "as is." Since the third and fourth terms are negative (or zero), they are added by complementing each bit, adding "1" to the LSB column, and sign extending with a leading "1". With these substitutions, the product is computed without any subtractions as

$$\begin{aligned} P = {} & a_{m-1}b_{n-1}2^{m+n-2} + \sum_{i=0}^{m-2}\sum_{j=0}^{n-2} a_i b_j 2^{i+j} + \sum_{i=0}^{m-2} \overline{b_{n-1}a_i} 2^{i+n-1} \\ & + \sum_{j=0}^{n-2} \overline{a_{m-1}b_j} 2^{j+m-1} + 2^{m+n-1} + 2^{n-1} + 2^{m-1} \end{aligned} \tag{3}$$

Figure 1 shows the multiplication of two 8-bit integers in two's complement form. The partial product bit matrix is described by Eq. (3),

$$
\begin{array}{cccccccccccccccc}
 & & & & & & & A & a_7 & a_6 & a_5 & a_4 & a_3 & a_2 & a_1 & a_0 \\
 & & & & & & & \times B & b_7 & b_6 & b_5 & b_4 & b_3 & b_2 & b_1 & b_0 \\
\hline
 & & & & & & & 1 & \overline{a_7b_0} & a_6b_0 & a_5b_0 & a_4b_0 & a_3b_0 & a_2b_0 & a_1b_0 & a_0b_0 \\
 & & & & & & & \overline{a_7b_1} & a_6b_1 & a_5b_1 & a_4b_1 & a_3b_1 & a_2b_1 & a_1b_1 & a_0b_1 & \\
 & & & & & & \overline{a_7b_2} & a_6b_2 & a_5b_2 & a_4b_2 & a_3b_2 & a_2b_2 & a_1b_2 & a_0b_2 & & \\
 & & & & & \overline{a_7b_3} & a_6b_3 & a_5b_3 & a_4b_3 & a_3b_3 & a_2b_3 & a_1b_3 & a_0b_3 & & & \\
 & & & & \overline{a_7b_4} & a_6b_4 & a_5b_4 & a_4b_4 & a_3b_4 & a_2b_4 & a_1b_4 & a_0b_4 & & & & \\
 & & & \overline{a_7b_5} & a_6b_5 & a_5b_5 & a_4b_5 & a_3b_5 & a_2b_5 & a_1b_5 & a_0b_5 & & & & & \\
 & & \overline{a_7b_6} & a_6b_6 & a_5b_6 & a_4b_6 & a_3b_6 & a_2b_6 & a_1b_6 & a_0b_6 & & & & & & \\
1 & a_7b_7 & \overline{a_6b_7} & \overline{a_5b_7} & \overline{a_4b_7} & \overline{a_3b_7} & \overline{a_2b_7} & \overline{a_1b_7} & \overline{a_0b_7} & & & & & & & \\
\hline
p_{15} & p_{14} & p_{13} & p_{12} & p_{11} & p_{10} & p_9 & p_8 & p_7 & p_6 & p_5 & p_4 & p_3 & p_2 & p_1 & p_0
\end{array}
$$

Figure 1 8×8 partial product bit matrix (two's complement).

and is implemented using an array of AND and NAND gates. The matrix is then reduced using techniques such as Wallace [6], Dadda [7], or reduced area reduction [8].

B. Truncated Multipliers

Truncated $m \times n$ multipliers, which produce results less than $m + n$ bits long, are described in [9]. Benefits of truncated multipliers include reduced area, delay, and power consumption [10]. An overview of truncated multipliers, which discusses several methods for correcting the error introduced due to unformed partial product bits, is given in [11]. The method used in this chapter is constant correction, as described in [9].

Figure 2 shows an 8×8 truncated parallel multiplier with a correction constant added. The final result is l-bits long. We define k as the number of truncated columns that are formed, and r as the number of columns that are not formed. In this example, the five least significant columns of partial product bits are not formed ($l = 8$, $k = 3$, $r = 5$).

Truncation saves an AND gate for each bit not formed and eliminates the full adders and half adders that would otherwise be required to reduce them to two rows. The delay due to reducing the partial product matrix is not improved because the height of the matrix is unchanged. However, a shorter carry propagate adder is required, which may improve the overall delay of the multiplier.

The correction constant, C_r, and the "1" added for rounding are normally included in the reduction matrix. In Fig. 2 they are explicitly shown to make the concept more clear.

A consequence of truncation is that a reduction error is introduced due to the discarded bits. For simplicity, the operands are assumed to be inte-

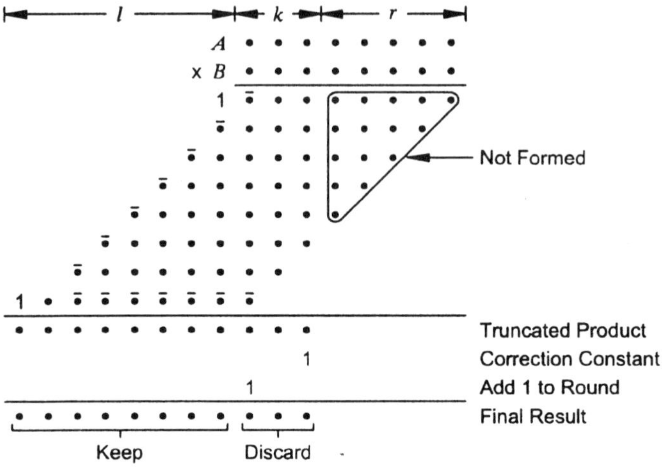

Figure 2 8 × 8 truncated multiplier with correction constant.

gers, but the technique can also be applied to fractional or mixed number systems. With r unformed columns, the reduction error is

$$E_r = -\sum_{i=0}^{r-1}\sum_{j=0}^{i} a_{i-j}b_j 2^i \tag{4}$$

If A and B are random with a uniform probability density, then the average value of each partial product bit is $\frac{1}{4}$, so the average reduction error is

$$\begin{aligned} E_{r\_avg} &= -\frac{1}{4}\sum_{q=0}^{r-1}(q+1)2^q \\ &= -\frac{1}{4}((r-1)\cdot 2^r + 1) \end{aligned} \tag{5}$$

The correction constant, C_r, is chosen to offset E_{r_avg} and is

$$\begin{aligned} Cr &= -\text{round}(2^{-r}E_{r\_avg})\cdot 2^r \\ &= -\text{round}\big((r-1)\cdot 2^{-2} + 2^{-(r+2)}\big)\cdot 2^r \end{aligned} \tag{6}$$

where round (x) indicates x is rounded to the nearest integer.

II. FIR FILTER ARCHITECTURE

This section describes the architecture used to study the effect of truncated multipliers in FIR filters. Little work has been published in this area, and this architecture incorporates the novel approach of combining all constants for two's complement multiplication and correction of reduction error into a single constant added just prior to computing the final filter output. This technique reduces the average reduction error of the filter by several orders of magnitude, when compared to the approach of including thc constants directly in the multipliers. Subsection A presents an overview of the architecture, and subsection B describes components within the architecture.

A. Architecture Overview

An FIR filter with T taps computes the following difference equation [12],

$$y[n] = \sum_{k=0}^{T-1} b[k] \cdot x[n-k] \tag{7}$$

where $x[\,]$ is the input data stream, $b[k]$ is the kth tap coefficient, and $y[\,]$ is the output data stream of the filter. Since the tap coefficients and the impulse response, $h[n]$, are related by

$$h[n] = \begin{cases} b[n], n = 0, 1, \ldots, T-1 \\ 0, \quad \text{otherwise} \end{cases} \tag{8}$$

Equation (7) can be recognized as the discrete convolution of the input stream with the impulse response [12].

Figure 3 shows the block diagram of the FIR filter architecture used in this chapter. This architecture has two data inputs, `x_in` and `coeff`, and one data output, `y_out`. There are two control inputs that are not shown, `clk` and `loadtap`.

The input data stream enters at the `x_in` port. When the filter is ready to process a new sample, the data at `x_in` is clocked into the register labeled $x[n]$ in the block diagram. The $x[n]$ register is one of T shift registers, where T is the number of taps in the filter. When `x_in` is clocked into the $x[n]$ register, the values in the other registers are shifted right in the diagram, with the oldest value, $x[n - T + 1]$ being discarded.

The tap coefficients are stored in another set of shift registers, labeled $b[0]$ through $b[T-1]$ in Fig. 3. Coefficients are loaded into the registers by applying the coefficient values to the `coeaff` port in sequence and cycling the `loadtap` signal to load each one.

The filter is pipelined with four stages: operand selection, multiplication, summation, and final addition.

Operand selection. The number of multipliers in the architecture is configurable. For a filter with T taps and M multipliers, each mul-

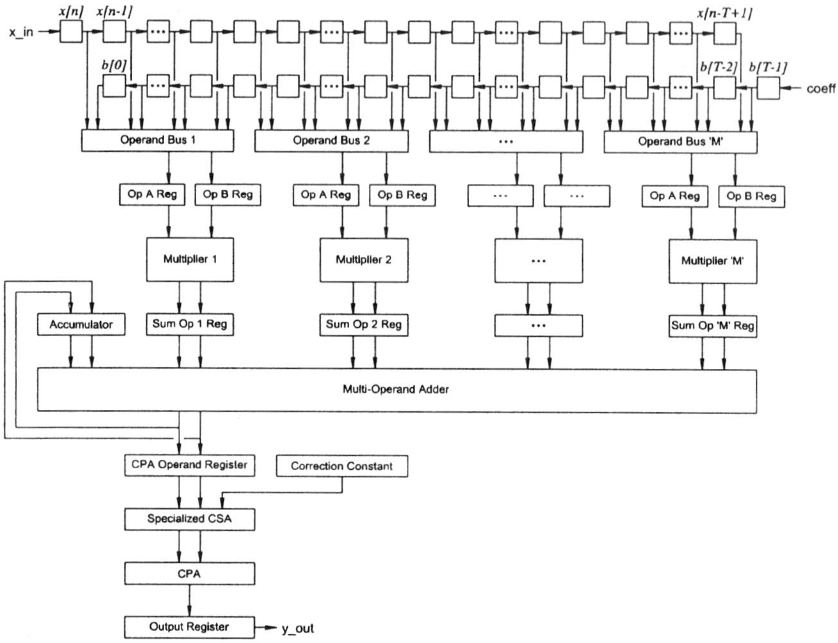

Figure 3 Proposed FIR filter architecture with T taps and M multipliers.

tiplier performs $\lceil T/M \rceil$ multiplications per input sample. The operands for each multiplier are selected each clock cycle by an operand bus and clocked into registers.

Multiplication. Each multiplier has two input operand registers, loaded by an operand bus in the previous stage. Each pair of operands is multiplied, and the final two rows of the reduction tree (the product in carry-save form) are clocked into a register where they become inputs to the multi-operand adder in the next stage. Keeping the result in carry-save form, rather than using a carry propagate adder (CPA), reduces the overall delay.

Summation. The multi-operand adder has carry-save inputs from each multiplier, as well as a carry-save input from the accumulator. After each of the $\lceil T/M \rceil$ multiplications have been performed, the output of the multi-operand adder (in carry-save form) is clocked into the CPA operand register where it is added in the next pipeline stage.

Final addition. In the final stage, the carry-save vectors from the multi-operand adder and a correction constant are added by a specialized carry-save adder and a carry-propagate adder to produce a single result vector. The result is then clocked into an output register, which is connected to the `y_out` output port of the filter.

The `clk` signal clocks the system. The clock period is set so that the multipliers and the multi-operand adder can complete their operation within one clock cycle. Therefore, $\lceil T/M \rceil$ clock cycles are required to process each input sample. The final addition stage only needs to operate once per input sample, so it has $\lceil T/M \rceil$ clock cycles to complete its calculation and is generally not on the critical path.

B. Architecture Components

This section discusses the components of the FIR filter architecture.

1. Multipliers

In this chapter, two's complement parallel tree multipliers are used to multiply the input data by the filter coefficients. When performing truncated multiplication, the constant correction method [9] is used. The output of each multiplier is the final two rows remaining after reduction of the partial product bits, which is the product in carry-save form [13]. Rounding does not occur at the multipliers, each product is $(l + k)$-bits long. Including the extra k bits in the summation avoids an accumulation of roundoff errors. Rounding is done in the final addition stage.

As described in subsection A, the last three terms in Eq. (3) are constants. In this architecture, these constants are *not* included in the partial product matrix. Likewise, if using truncated multipliers, the correction constant is not included either. Instead, the constants for each multiplication are added in a single operation in the final addition stage of the filter. This is described later in more detail.

2. Multi-Operand Adder and Accumulator

As shown in Eq. (7), the output of an FIR filter is a sum of products. In this architecture, M products are computed per clock cycle. In each clock cycle, the carry-save outputs of each multiplier are added and stored in the accumulator register, also in carry-save form. The accumulator is included in the sum, except with the first group of products for a new input sample. This is accomplished by clearing the accumulator when the first group of products arrives at the input to the multi-operand adder.

The multi-operand adder is simply a counter reduction tree, similar to a counter reduction tree for a multiplier, except that it begins with operand bits from each input instead of a partial product bit matrix. The output of the multi-operand adder is the final two rows of bits remaining after reduction, which is the sum in carry-save form. This output is clocked into the accumulator register every clock cycle, and clocked into the carry propagate adder (CPA) operand register every $[T/M]$ cycles.

3. Correction Constant Adder

As stated previously, the constants required for two's complement multipliers and the correction constant for unformed bits in truncated multipliers are not included in the reduction tree but are added during the final addition stage. A "1" for rounding the filter output is also added in this stage. All of these constants for each multiplier are precomputed and added as a single constant, C_{TOTAL}.

All multipliers used in this chapter operate on two's complement operands. From Eq. (3), the constant that must be added for an $m \times n$ multiplier is $2^{m+n-1} + 2^{n-1} + 2^{m-1}$. With T taps, there are T multiply operations (assuming T is evenly divisible by M), so a value of

$$C_M = T(2^{m+n-1} + 2^{n-1} + 2^{m-1}) \tag{9}$$

must be added in the final addition stage.

The multipliers may be truncated with unformed columns of partial product bits. If there are unformed bits, the total average reduction error of the filter is $T \cdot E_{r\_avg}$. The correction for this is

$$C_R = \text{round}\left(T \cdot (r-1) \cdot 2^{-2} + T \cdot 2^{-(r+2)}\right) \cdot 2^r \tag{10}$$

To round the filter output to l bits, the rounding constant that must be used is

$$C_{RND} = 2^{r+k-1} \tag{11}$$

Combining these constants, the total correction constant for the filter is

$$C_{TOTAL} = C_M + C_R + C_{RND} \tag{12}$$

Adding C_{TOTAL} to the multi-operand adder output is done using a specialized carry-save adder (SCSA), which is simply a carry-save adder optimized for adding a constant bit vector. A carry-save adder uses full adders to reduce three bit vectors to two. SCSAs differ in that half adders are used in columns where the constant is a "0" and specialized half adders are used in columns where the constant is a "1". A specialized half adder computes the sum and carry-out of two bits plus a "1", the logic equations being

$$s_i = \overline{a_i \oplus b_i} \qquad \text{and} \qquad c_{i+1} = a_i + b_i \tag{13}$$

The output of the SCSA is then input to the final carry propagate adder.

4. Final Carry Propagate Adder

The output of the specialized carry-save adder is the filter output in carry-save form. A final CPA is required to compute the final result. The final addition stage has $\lceil T/M \rceil$ clock cycles to complete, so for many applications a simple ripple-carry adder will be fast enough. If additional performance is required, a carry-lookahead adder may be used. Using a faster CPA does not increase throughput, but does improve latency.

5. Control

A filter with T taps and M multipliers requires $\lceil T/M \rceil$ clock cycles to process each input sample. The control circuit is a state machine with $\lceil T/M \rceil$ states, implemented using a modulo-$\lceil T/M \rceil$ counter. The present state is the output of the counter and is used to control which operands are selected by each operand bus. In addition to the present state, the control circuit generates four other signals: (1) `shiftData`, which shifts the input samples (2) `clearAccum`, which clears the accumulator, (3) `loadCpaReg`, which loads the output of the multi-operand adder into the CPA operand register, and (4) `loadOutput`, which loads the final sum into the output register.

III. FILTER GENERATION SOFTWARE

The architecture described in Section II provides a great deal of flexibility in terms of operand size, the number of taps, and the type of multipliers used. This implies that the design space is quite large. In order to facilitate the development of a large number of specific implementations, a tool was designed that automatically generates synthesizable structural VHDL models given a set of parameters. The tool, which is named filter generation software (FGS), also generates test benches and files of test vectors to verify the filter models.

FGS is written in Java and consists of two main packages. The arithmetic package, discussed in subsection A, is suitable for general use and is the foundation of FGS. The fgs package, discussed in subsection B, is specifically for generating the filters described previously. It uses the arithmetic package to generate the necessary components.

A. The arithmetic Package

The arithmetic package includes classes for modeling and simulating digital components. The simplest components include D flip-flops, half adders, and full adders. Larger components such as ripple-carry adders and parallel multipliers use the smaller components as building blocks. These components, in turn, are used to model complex systems such as FIR filters.

1. Common Classes and Interfaces

Figure 4 shows the classes and interfaces used by arithmetic subpackages. The most significant are VHDLGenerator, Parameterized, and Simulator.

VHDLGenerator is an abstract class. Any class that represents a digital component and can generate a VHDL model of itself is derived from this class. It defines three abstract methods that must be implemented by all subclasses. genCompleteVHDL() generates a complete VHDL file describing the component. This file includes synthesizable entity-architecture descriptions of all subcomponents used. genComponentDeclaration() generates the component declaration that must be included in the entity-architecture descriptions of other components that use this component. genEntityArchitecture() generates the entity-architecture description of this component.

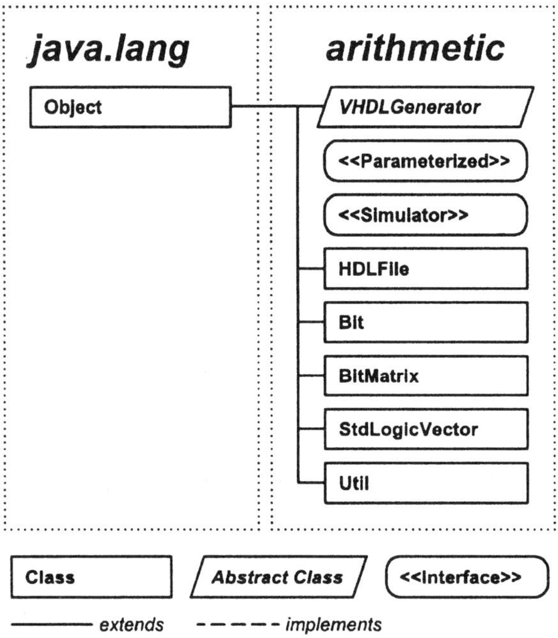

Figure 4 The arithmetic package.

Parameterized is an interface implemented by classes whose instances can be defined by a set of parameters. The interface includes get and set methods to access those parameters. Specific instances of Parameterized components can be easily modified by changing these parameters.

Simulator is an interface implemented by classes that can simulate their operation. The interface has only one method, simulate, which accepts a vector of inputs and return a vector of outputs. These inputs and outputs are vectors of IEEE VHDL std_logic_vectors [14].

2. The arithmetic.smallcomponents Package

The arithmetic.smallcomponents package provides components such as D flip-flops and full adders that are used as building blocks for larger components such as registers, adders, and multipliers. Each class in this package is derived from VHDLGenerator enabling each to generate VHDL for use in larger components.

3. The arithmetic.adders Package

The classes in this package model various types of adders including carry-propagate adders, specialized carry-save adders, and multi-operand adders. All components in these classes handle operands of arbitrary length and weight. This flexibility makes automatic VHDL generation more complex than it would be if operands were constrained to be the same length and weight. However, this flexibility is often required when an adder is used with another component such as a multiplier.

Figure 5 shows the arithmetic.adders package, which is typical of many of the arithmetic subpackages. CarryPropagateAdder is an abstract class from which carry-propagate adders such as ripple-carry adders and carry-lookahead adders are derived. CarryPropagateAdder is a subclass of VHDLGenerator and implements the Simulator and Parameterized interfaces. Using interfaces and an inheritance hierarchy such as this help make FGS both straightforward to use and easy to extend. For example, a new type of carry-propagate adder could be incorporated into existing complex models by subclassing CarryPropagateAdder.

4. The arithmetic.matrixreduction Package

This package provides classes that perform matrix reduction, typically used by multi-operand adders and parallel multipliers. These classes perform Wallace, Dadda, and reduced area reduction [6–8]. Each of these classes are derived from the ReductionTree class.

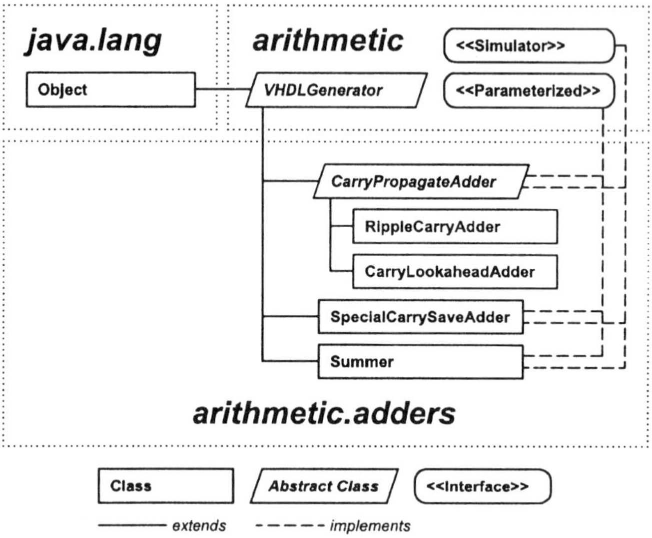

Figure 5 The arithmetic.adders package.

5. The arithmetic.multipliers Package

A ParallelMultiplier class was implemented for this chapter and is representative of how FGS functions.

Parameters can be set to configure the multiplier for unsigned, two's complement, or combined operation. The number of unformed columns, if any, and the type of reduction, Wallace, Dadda, or reduced area, may also be specified. A BitMatrix object, which models the partial product matrix, is then instantiated and passed to a ReductionTree object for reduction. Through polymorphism (dynamic binding), the appropriate subclass of ReductionTree reduces the BitMatrix to two rows. These two rows can then be passed to a CarryPropagateAdder object for final addition, or in the case of the FIR filter architecture described in this chapter, to a multi-operand adder.

The architecture of FGS makes it easy to change the bit matrix, reduction scheme, and final addition methods. New techniques can be added seamlessly by subclassing appropriate abstract classes.

6. The arithmetic.misccomponents Package

This package includes classes that provide essential functionality but do not logically belong in other packages. This includes Bus, which models the operand busses of the FIR filter, and Register, which models various types of data registers. Implementation of registers is done by changing the type of flip-flop objects that comprise the register.

7. The arithmetic.firfilters Package

This package includes classes for modeling ideal FIR filters as well as FIR filters based on the truncated architecture described in Section II.

The "ideal" filters are ideal in the sense that the data and tap coefficients are double precision floating point. This is a reasonable approximation of infinite precision for most practical applications. The purpose of an ideal FIR filter object is to provide a baseline for comparison with practical FIR filters and to allow measurement of calculation errors.

The FIRFilter class models FIR filters based on the architecture shown in Fig. 3. All operands in FIRFilter objects are considered to be two's complement integers, and the multipliers and the multi-operand adder use reduced area reduction. There are many parameters that can be set including the tap coefficient and data lengths, the number of taps, the number of multipliers, and the number of unformed columns in the multipliers.

8. The arithmetic.testing Package

This package provides classes for testing components generated by other classes, including parallel multipliers and FIR filters. The FIR filter test class generates a test bench and an input file of test vectors. It also generates a.vec file for simulation using Altera Max+Plus II.

9. The arithmetic.gui Package

This package provides graphical user interface (GUI) components for setting parameters and generating VHDL models for all of the larger components such as FIRFilter, ParallelMultiplier, etc. The GUI for each component is a Java Swing JPanel, which can be used in any swing application. These panels make setting component parameters and generating VHDL files simple and convenient.

B. The fgs Package

Whereas the arithmetic package is suitable for general use, the fgs package is specific to the FIR filter architecture described in Section II. fgs includes classes for automating much of the work done to analyze the use of truncated multipliers in FIR filters. For example, this package includes a driver class that automatically generates a large number of different FIR filter configurations for synthesis and testing. Complete VHDL models are then generated, as well as Tcl scripts to drive the synthesis tool. The Tcl script commands the synthesis program to write area and delay reports to disk files, which are parsed by another class in the fgs package that summarizes the data and writes it to a CSV file for analysis by a spreadsheet application.

IV. RESULTS

Table 1 presents some representative synthesis results that were obtained from the Leonardo synthesis tool and the LCA300K 0.6 micron CMOS

Table 1 Synthesis Results for Filters with 16-bit Operands, Output Rounded to 16-bits (Optimized for Area)

Filter			Synthesis results			Improvement (%)		
T	M	r	Area (gates)	Total delay (ns)	$A{\cdot}D$ product (gates·ns)	Area	Total delay	A·D product
12	2	0	16241	40.80	662633	—	—	—
12	2	12	12437	40.68	505937	23.4	0.3	23.6
12	2	16	10211	40.08	409257	37.1	1.8	38.2
16	2	0	17369	54.40	944874	—	—	—
16	2	12	13529	54.24	733813	22.1	0.3	22.3
16	2	16	11303	53.44	604032	34.9	1.8	36.1
20	2	0	19278	68.00	1310904	—	—	—
20	2	12	15475	67.80	1049205	19.7	0.3	20.0
20	2	16	13249	66.80	885033	31.3	1.8	32.5
24	2	0	20828	81.60	1699565	—	—	—
24	2	12	17007	81.36	1383690	18.3	0.3	18.6
24	2	16	14781	80.16	1184845	29.0	1.8	30.3
12	4	0	25355	20.40	517242	—	—	—
12	4	12	18671	20.34	379768	26.4	0.3	26.6
12	4	16	14521	20.04	291001	42.7	1.8	43.7
16	4	0	26133	27.20	710818	—	—	—
16	4	12	19413	27.12	526481	25.7	0.3	25.9
16	4	16	15264	26.72	407854	41.6	1.8	42.6
20	4	0	28468	34.00	967912	—	—	—
20	4	12	21786	33.90	738545	23.5	0.3	23.7
20	4	16	17636	33.40	589042	38.0	1.8	39.1
24	4	0	29802	40.80	1215922	—	—	—
24	4	12	23101	40.68	939749	22.5	0.3	22.7
24	4	16	18950	40.08	759516	36.4	1.8	37.5

standard cell library. Improvements in area, delay, and area-delay product for filters using truncated multipliers are given relative to comparable filters using standard multipliers. Table 2 presents reduction error figures for 16-bit filters with T taps and r unformed columns. Additional data can be found in [15], which also provides a more detailed analysis of the FIR filter architecture presented in this chapter, including reduction and roundoff error. The main findings were:

1. Using truncated multipliers in FIR filters results in significant improvements in area. For example, the area of a 16-bit filter with 4 multipliers and 24 taps improves by 22.5% with 12 unformed columns and by 36.4% with 16 unformed columns. We estimate substantial power savings would be realized as well. Truncation has little impact on the overall delay of the filter.
2. The computational error introduced by truncation is tolerable for many applications. For example, the reduction error SNR for a 16-bit filter with 24 taps is 86.7 dB with 12 unformed columns and 61.2dB with 16 unformed columns. In comparison, the roundoff error for an equivalent filter without truncation is 89.1dB [15].
3. The average reduction error of a filter is independent of r (for T > 4), and much less than that of a single truncated multiplier. For a 16-bit filter with 24 taps and r = 12, the average reduction error is only 9.18×10^{-5} ulps, where an ulp is a unit of least

Table 2 Reduction Error for Filters with 16-bit Operands, Output Rounded to 16-bits

Filter		Reduction error		
T	r	SNR_R (dB)	σ_R (ulps)	E_{AVG} (ulps)
12	0	∞	0	0
12	12	89.70	0.268	−4.57E-5
12	16	64.22	5.040	−4.57E-5
16	0	∞	0	0
16	12	88.45	0.310	−6.10E-5
16	16	62.97	5.820	−6.10E-5
20	0	∞	0	0
20	12	87.48	0.346	−7.60E-5
20	16	62.00	6.508	−7.60E-5
24	0	∞	0	0
24	12	86.69	0.379	−9.18E-5
24	16	61.21	7.143	−9.18E-5

precision in the 16-bit product. In comparison, the average reduction error of a single 16-bit multiplier with r = 12 is 1.56×10^{-2} ulps, and the average roundoff error of the same multiplier without truncation is 7.63×10^{-6} ulps.

V. CONCLUSIONS

This chapter presented a tool used to rapidly prototype parameterized FIR filters. The tool was used to study the effects of using truncated multipliers in those filters. It was based on a package of arithmetic classes that are used as components in hierarchical designs, and are capable of generating structural level VHDL models of themselves. Using these classes as building blocks, FirFilter objects generate complete VHDL models of specific FIR filters. The arithmetic package is extendable and suitable for use in other applications, enabling rapid prototyping of other computational systems. As a part of ongoing research at Lehigh University, the tool is being expanded to study other DSP applications, and will be made available to the public in the near future.*

REFERENCES

1. Lightbody, G., Walke, R., Woods, R. F., McCanny, J. V. (1998). Rapid System Prototyping of a Single Chip Adaptive Beamformer. In: Proceedings of Signal Processing Systems, 285–294.
2. McCanny, J., Ridge, D., Yi, H., Hunter, J. (1997). Hierarchical VHDL Libraries for DSP ASIC Design. Proceedings of the IEEE International Conference on Acoustics, Speech, and Signal Processing, 675–678.
3. Pihl, J., Aas, E. J. (1996). A Multiplier and Squarer Generator for High Performance DSP Applications. In: Proceedings of the 39th Midwest Symposium on Circuits and Systems 109–112.
4. Richards, M. A., Gradient, A. J., Frank, G. A. (1997). *Rapid Prototyping of Application Specific Signal Processors*. Kluwer Academic Publishers.
5. Saultz, J. E. (1997). Rapid Prototyping of Application-Specific Signal Processors (RASSP) In-Progress Report. *Journal of VLSI Signal Processing Systems for Signal, Image, and Video Technology* 29–47.
6. Wallace, C.S. (1964). A Suggestion for a Fast Multiplier. *IEEE Transactions on Electronic Computers EC-13* 14–17.

*As of this writing, the software has been refactored and given the name "Magellan." The current state of Magellan can be found at http://www.lehigh.edu/~gew5/magellan.html, or by contacting the authors.

7. Dadda, L. (1965). Some Schemes for Parallel Multipliers. *Alta Frequenza* 34: 349–356.
8. Bickerstaff, K. C., Schulte, M. J., Swartzlander, E. E. (1995). Parallel Reduced Area Multipliers. *IEEE Journal of VLSI Signal Processing* 9:181–191.
9. Schulte, M. J., Swartzlander, E. E. (1993). Truncated Multiplication with Correction Constant. *VLSI Signal Processing VI*. Eindhoven, Netherlands: IEEE Press, pp. 338–396.
10. Schulte, M. J., Stine, J. E., Jansen, J. G. (1999). Reduced Power Dissipation Through Truncated Multiplication. IEEE Alessandro Volta Memorial Workshop on Low Power Design, Como, Italy, 61–69.
11. Swartzlander, E. E. (1999). Truncated Multiplication with Approximate Rounding. In: Proceedings of the 33rd Asilomar Conference on Signals, Circuits, and Systems, 1480–1483.
12. Oppenheim, A. V., Schafer, R. W. (1999). *Discrete-Time Signal Processing*. 2nd ed. Upper Saddle River, NJ: Prentice Hall.
13. Koren, I. (1993). *Computer Arithmetic and Algorithms*. Englewood Cliffs, NJ: Prentice Hall.
14. IEEE Standard Multivalue Logic System for VHDL Model Interoperability (Stdlogic1164): IEEE Std 1164–1993 (26 May 1993).
15. Walters, E.G. III (2002). Design Tradeoffs Using Truncated Multipliers in FIR Filter Implementations. Master's thesis, Lehigh University.

2

An LUT-Based High Level Synthesis Framework for Reconfigurable Architectures

Loïc Lagadec, Bernard Pottier, and Oscar Villellas-Guillen
Université de Bretagne Occidentale, UFR Sciences, Brest, France

I. INTRODUCTION

A. General Context

It is a fact that integration technology is providing hardware resources at an exponential rate while the development methods in industry are only progressing at a linear rate. This can be seen as the repetition of a common situation where a mature technical knowledge is providing useful possibilities in excess of current method capabilities. An answer to this situation is to change the development process in order to avoid work repetition and to provide more productivity by secure assembly of standard components.

This situation is also known from computer scientists since it was encountered earlier in the programming language story [1]. The beginnings of this story were: (1) symbolic expression of computations (Fortran), (2) structured programs (Algol), (3) modularity, and code abstraction via interfaces and hiding (Modula2, object-oriented programming).

Modularity came at an age when efficient engineering of large programs was the main concern, and when the task of programming was overtaken by organizational problems. System development can be considered as a new age for computer architecture design, with hardware description languages needing to be transcended by a higher level of description to increase productivity. Companies developing applications have specific methods for design and production management in which they can represent their products, tools, and hardware or software components. The method that ensures the feasibility of a product leads technical choices and developments. It also changes some of the rules in design organization, since most applications are achieved in a top-down fashion using object models or code generators to reach functional requirements.

B. Reconfigurable Architectures

FPGAs are one of the driving forces for integration technology progresses, due to their increasing applications. Like software and hardware programming languages, reconfigurable architectures are sensitive to scale mutations. As the chip size increases, the characteristics of the application architecture change with new needs for structured communications, more efficiency on arithmetic operators, and partial reconfigurability.

The software follows slowly, migrating from HDL (Hardware Description Language) to HLL (High Level Language). Preserving the developments and providing a sane support for production tools is a major issue. Reconfigurable architectures can take benefits from the top-down design style of subsection A, by a high level of specialization in applications.

C. Madeo

MADEO is a medium term project that makes use of open object modeling to provide a portable access to hardware resources and tools on reconfigurable architectures.

The project structure has three parts that interact closely (bottom-up):

1. *Reconfigurable architecture model and its associated generic tools.* The representation of practical architectures on a generic model enables sharing of basic tools such as place and route, allocation, circuit edition [2]. Mapping a logic description to a particular technology is achieved using generic algorithms from SIS [3], or PPart [4]. Specific atomic resources such as memories, sensors or operators, can be merged with logic, and the framework is extensible.
2. *High-level logic compiler.* This compiler produces circuits associated to high level functionalities on a characterization of the above model. Object-oriented programming is not restricted to a particular set of operators or types, and then provides the capability to produce primitives for arbitrary arithmetics or symbolic computing.

   The compiler handles an intermediate level, which is a graph of lookup-tables carrying high-level values (objects). Then this graph is translated into a logic graph that will be mapped on hardware resources. The translator makes use of values produced in the high-level environment, which allows implementation of classical optimizations without attaching semantics to operations at the language level.

3. *System and architecture modeling*. The computation architecture in its static or dynamic aspects is described in this framework. For instance, these are generic regular architectures with their associated tools, processes, platform management, and system activity.

   The compiler can make use of logic generation to produce configurations, bind them to registers or memories, and produce a configured application. The ability to control placing and routing given by the first part, and synthesis from the second part, allows building of complex networks of fine- or medium-grain elements.

This chapter focuses on the logic compiler. Historically, this work has taken ideas from symbolic translation to logic as described in [5] and knowledge on automatic generation of primitives in interpreters [6]. Relation to the object-oriented environment was described in [7] with limited synthesis capabilities that are removed in current work. System modeling and program synthesis has been demonstrated in the case study of a smart sensor camera [8] based on the same specification syntax as the one used in the current work.

This chapter describes the general principles used for specification and logic production, then details the transformations that are achieved. An illustration is given with an example of a coder/decoder family for RAID (Redundant Array of Inexpensive Disks) systems with quantitative results in Section V.

II. A FRAMEWORK FOR LOGIC SYNTHESIS

A. Architecture Modeling

Reconfigurable architectures can mix different grains of hardware resources: logic Elements, operators, communication lines, buses, switches, memories, processors, and so on.

Most FPGAs (Field Programmable Gate Arrays) provide logic functions using small lookup memories (LUT) addressed by a set of signals. As seen from the logic synthesis tools, an n-bit wide LUT is the most general way to produce any logic function of n boolean variables. There are known algorithms and tools for partitioning large logic tables or networks to target a particular LUT-based architecture.

LUTs are effectively interconnected during the configuration phase to form logic. This is achieved using various configurable devices such as programmable interconnect points, switches, or shared lines. Some commercial architectures also group several LUTs and registers into cells called configurable logic blocks (CLB).

Our model for the organization of these architectures is a hierarchy of geometric patterns of hardware resources. The model is addressed via a specific grammar [2] allowing the description of concrete architectures. Given this description, generic tools operate for technology mapping, and placing and routing logic modules. See Figure 9 shows a view of the generic editor. Circuits such as operators or computing networks are described by programs realizing the geometric assembly of such modules and their connections.

Using this framework, few days of work are sufficient to bring up the set of tools on a new architecture with the possibility to port application components. On a concrete platform, it is then necessary to build the bitstream generation software by rewriting the application descriptions to the native tools. Two practical examples are the xc6200 that has a public architecture and has been addressed directly, the Virtex 1 is addressed through the JBits API, and other implementations include industrial prototype architectures.

If behavioral generators are known to offer numerous benefits over HDL synthesis, including ease of specifying a specialized design and the ability to perform partial evaluation [9], they generally remain dependent of some libraries of modules appearing as primitives [10,11]. Our approach draws attention to itself by relying on a generic back-end tool in charge of the modules production. There are no commercial tools or library involved in the flow.

B. Programming Considerations

Applications for fine-grain reconfigurable architectures can be specialized without compromise and they should be optimized in terms of space and performance. In our view, too much emphasis is placed on the local performance of standard arithmetic units in the synthesis tools and also in the specification languages.

A first consequence of this advantage is the restricted range of basic types coming from the capabilities of ALU/FPUs or memory address mechanisms. Control structures strictly oriented toward sequentiality are another aspect that can be criticized. As an example, programming for multimedia processor accelerators remains procedural in spite of all the experiences available from the domain of data parallel languages. Hardware description languages have rich descriptive capabilities, however, the necessity to use libraries has led the language designers to restrict their primitives to a level similar to **C**.

Our aim was to produce a more flexible specification level with direct and efficient coupling to logic. This implies allowing easy creation of specific arithmetics representing the algorithm needs, letting the compilers automatically tune data width, and modeling computations based on well-understood object classes.

The expected effect was an easy production of dedicated support for processes that need a high-level of availability, or would waste processor resources in an integrated system. To reach this goal, we used specifications with symbolic and functional characteristics with separate definition of data on which the program is to operate.

Sequential computations can be structured in various ways by splitting programs on register transfers, either explicitly in the case of an architecture description, or implicitly during the compilation. Figure 1 shows these two aspects, with a circuit module assembled in a pipeline and in a data-path. In the case of simple control loops or state machines, high-level variables can be used to retain the initial state with known values with the compiler retrieving progressively the other states by enumeration [7]. Figure 2 shows a diagram where registers are provided to hold state values associated to high-level variables that could be instance variables in an object.

At this stage, we will consider the case of methods without side-effects operating on a set of objects. For the sake of simplicity we will rename these

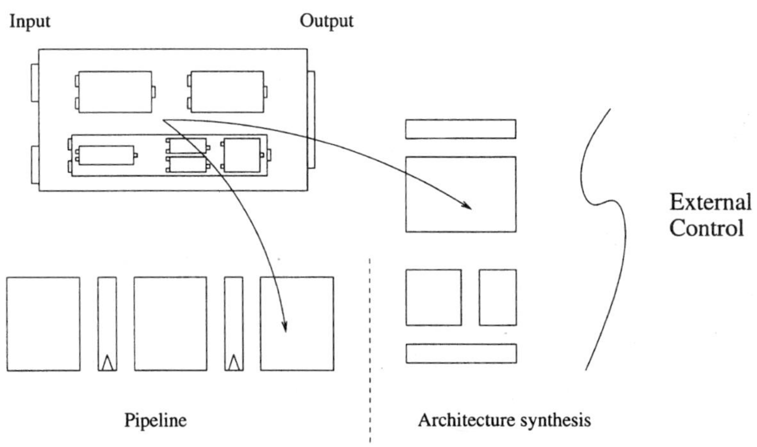

Figure 1 The modules can be either flat or hierarchical. The modules can be composed in order to produce pipelines or can be instantiated during architecture synthesis.

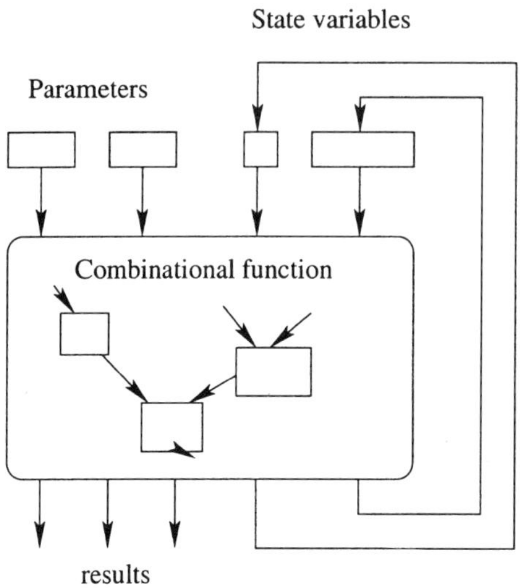

Figure 2 State machines can be obtained by methods operating on private variables having known initial values.

methods *functions*, and the set of objects values. Interaction with external variables is not discussed there. The input language is Smalltalk-80, and variant VisualWorks, which is also used to build the tools and to describe the application architectures.

C. Execution Model

The execution model targeted by the compiler is currently a high-level replication of LUT-based FPGAs. We define a *program* as a function that needs to be executed on a set of input values. Thus the notion of program groups the algorithm and the data description at once. Our *program* can be embedded in higher level computations of various kind, implying variables or memories. Data descriptions are inferred from these levels. The resulting circuit is highly dependent from the data it is intended to process.

An execution is the traversal of a hierarchical network of lookup tables in which values are forwarded. A value change in the input of a table implies a possible change in its output that in turn induces other changes downstream. These networks reflect the effective function structure at the procedure call grain and they have a strong algorithmic meaning. Among the different possibilities offered for practical execution, there are cascaded hash table accesses and use of general purpose arithmetic units where they are detected to fit.

Translation to FPGAs need binary representation for objects, as shown in Fig. 6. This is achieved in two ways, by using a specific encoding known to be efficient, or by exchanging object values appearing in the input and output for indexes in the enumeration of values. Figure 3 shows a fan-in case with an aggregation of indexes in the input of function *h*(). Basically the low-level representation of a node such as *h*() is a programmable logic array (PLA), having in its input the Cartesian product of the set of incoming indexes (*fout* × *gout*), and in its output the set of indexes for downstream.

Some important results or observations from this exchange are:

1. Data paths inside the network do not depend anymore on data width but on the *number of different values present on the edges.*
2. Depending on the interfacing requirements, it will be needed to insert nodes in the input and output of the network to handle the exchanges between values and indexes.
3. Logic synthesis tool capabilities are limited to medium-grain problems. To allow compilation to FPGAs, algorithms must *decrease the number of values* down to nodes that can be easily handled by the bottom layer (SIS partitioning for LUT-n). Today, this grain is similar to algorithms coded for 8-bit microprocessors.
4. *Decreasing the number of values* is the natural way in which functions operate, since the size of a Cartesian product on a function input values is the maximum number of values produced in the output. The number of values carried by edges decreases either

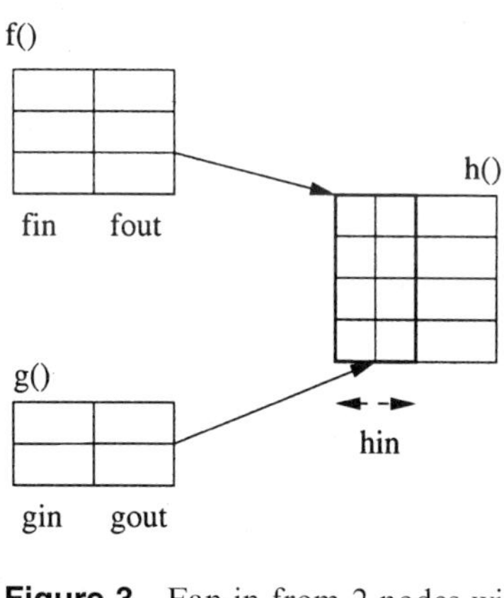

Figure 3 Fan-in from 2 nodes with *Card*(*fout* × *gout*) <*Card*(*fin*) × *Card*(*gin*).

in the hierarchy structure or in a graph flow. There is no possible divergence and the efficiency of an algorithm can be stated to be its ability to quickly decrease the data amplitude on which the logic complexity depends.

D. Type System

Language types appear to the programmers as annotations for checking code consistency and binding to architecture resources. The system type we are using does not restrict programming to this kind of binding. It is only intended to specify any possible set of values appearing in the program input or inside the computation network. In the object environment, it is supported by a set of classes supporting operations.

Implicit or explicit collections of values are denoted by intervals or sets. *Class-based types* are associated either to classes having a finite number of instances (booleans, bytes, or small integers), or to user-defined new functionalities, including arithmetics. Unions result from operations on the two previous types.

III. COMPILER FLOW

A. Flat Expressions

In the first stage, let us consider a program where the number of values appearing in the input of each function call is compatible with an efficient logic synthesis for an LUT-n FPGA architecture. As each node can be directly synthesized, we have a *flat* expression in opposition with *hierarchical* expressions that will need additional compilation contexts for some of the function calls.

As a Smalltalk development environment was used, there was an obvious interest to use this language syntax for *programs* targeting FPGAs. Immediate benefits were the reuse of the standard compiler front-end use of the existing classes.

1. Building the Value Network

The first compilation stage consists in building an acyclic flat graph, which nodes are lookup tables based on objects and which edges allow values to pass downstream.

As stated, the syntax tree is produced by the standard compiler. The directed acyclic graph (DAG) is built by analyzing the syntax tree and variable use. Local variable references are eliminated. At this stage nodes are still

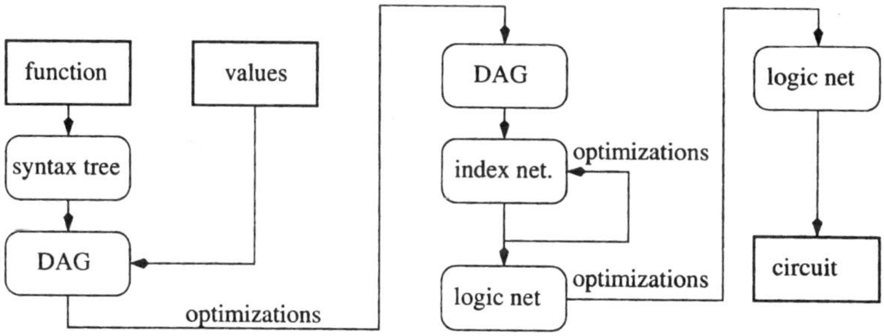

Figure 4 Compiler flow.

holding function calls that receive edges from the function parameter list, or other nodes.

To replace these nodes by lookup tables, the values are propagated progressively from the function parameter list. A graph traversal is achieved, building a table for each node with defined inputs.

During this transformation, care must be taken for dependencies in variable used in fan-out to fan-in subnets. For example, the composition $h[f(x,y), g(x,z)]$ has a smaller output than $h[f(x,y), g(t,z)]$ because of the dependency on variable x. A number of inputs in the fan-in node h and upstream are not useful and can be deleted by constraining the Cartesian products from f and g tables. A lot of conditional computations fall in this case.

2. High Level Optimization and Building the Index Network

After the first stage we have a situation similar to a compiler that has a language semantic knowledge because the tables have stronger properties inferred from the message executions. It is time to apply high-level optimizations such as elimination of constant nodes and dead code or subexpression factorization. This implies backward and forward processing on the DAG. Immediate benefits were the reuse of the standard compiler front-end. The compiler flow is shown in Figure 4.

The next transformation is the translation of the DAG by deducing index-based tables from associations of value tables. This is achieved by generating an index for values. Care must be taken for class-based types to preserve their special encoding.

3. LUT-Based Optimizations and Architecture Mapping

Index path optimizations involve the detection of subnets with particular topologies. For example, linear cascade tables can be collapsed into a single table. For logic translation, each index-based table is given logic synthesis tools to produce an equivalent binary description. At this stage we must also take into account the size of LUT memories in the target architecture. The result is a hierarchical logic description that is a binary equivalent for the high-level program.

The last stage is to place and route the logic graph using the generic tools in the framework to produce a hardware module for further system handling and binding.

B. Hierarchical Aspects

In subsection A we supposed that the program could be directly synthesized at each function call. We now consider the more general case where calls must be developed to reach this condition.

The logic needed to implement a particular function call depends on the expressed algorithm, the number of parameters, the number of possible values for parameters, and the original encoding of values in the higher level environment. A valuable property of an algorithm is its ability to quickly decrease the number of values present on graph edges. This gradual decrease comes from function calls that are processed in the same way of their root function with every node showing an excessive complexity related to synthesis.

When the compiler reaches a condition where logic tools will be inefficient, it creates a new compilation context and processes recursively the call. The context will return a structured logic description that will be installed as part of the current level production.

The technical form of a logic description associated to a compiled program is a hierarchical Berkeley logic interchange format (BLIF) description that can be partially flattened for further logic optimization, and partially placed under control of a floor planner. In this case each developed function call has its corresponding circuit component assembled in the global hierarchy.

A more speculative compiler built-in function is type partitioning. When a data set appears to be much too large, the compiler can divide the type in order to reach a grain suitable for synthesis. Automatic type division by the compiler should be considered only as a quick approximation, since the function algorithms are normally written to manage synthesis complexity at a high level.

A similar situation is the knowledge of a "best encoding" for values. For example, the order of elements in a Galois field has an influence on the logic complexity of basic operators. If these operations are dominant in the code, type-based rules must be managed by the compiler to prevent a new type generation in node outputs.

IV. EXAMPLE OF SMALL FP MULTIPLIERS

To explain the programming interface related to the compiler work, let us comment on the case of a multiplier operating on small floating-point numbers. The numbers are represented as a sign, an exponent, and a fractional part. The multiplication is described in a class supporting secondary methods for adding exponents, multiplying fractional parts, and normalizing the result. The algorithm for the multiplication is taken from [12]. It was developed and verified in the software environment before synthesis. Below is a Smalltalk-80 code for this.

```
sign: signA significand: significandA
exponent: exponentA
sign: signB significand: significandB
exponent: exponentB
| sign exp mant shift |
sign := self computeSignFor:signA and: signB.
mant := self computeSignificandsFor:
significandA and: significandB.
shift := self computeMantOffset: mant.
exp := (self computeExponentFor: exponentA and: expo-
nentB) + shift.
mant := self normalizeMant: mant with: shift.
¯Array with: sign with: mant with: exp
```

To produce logic for this program, it is necessary to provide a characterization of the objects present in the various fields as boolean or integer intervals, thus tailoring the arithmetic. Depending on the data amplitude of the inputs of each node, the compiler either develops them hierarchically or not. In the first case a new graph is produced which is reachable from the node. In the second case a table, equivalent to the node, will be directly computed and associated to the node.

Figure 5 shows the multiplier directed acyclic graph (DAG). Light grey boxes are associated to hierarchical nodes expansing their own DAG, dark grey boxes are hierarchical nodes flattened due to their small size. Medium grey boxes are associated to terminal primitives translated directly into lookup tables.

Figure 6 shows two different aspects of a particular node. The first table is holding high-level objects, such as rational numbers and the right view displays the equivalent index table with symbol numbers.

Figure 7 displays a 6-node hierarchical circuit and a flat globally optimized circuit for the multiplier. In the first case, there was no attempt to compact the design nor to achieve a flat logic optimization to help readability.

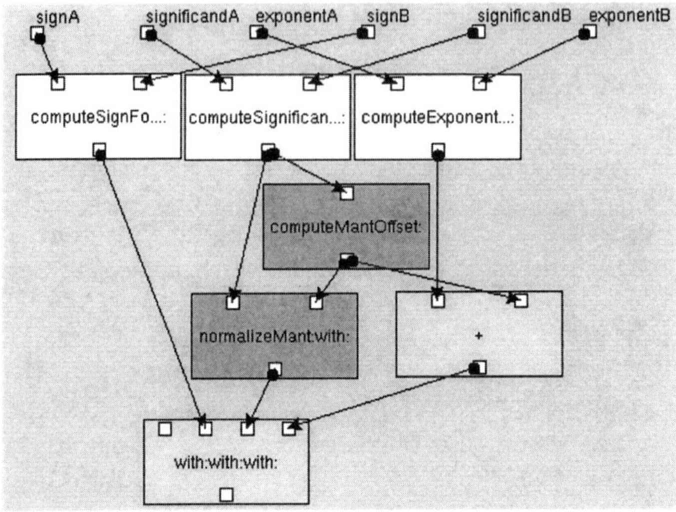

Figure 5 Data-flow graph for the multiplication after removing temporaries and complexity evaluation.

The technology is LUT4 and the circuits have, respectively, 161 and 138 nodes. The second circuit had 24 nets unrouted due to a choice of low routability in the FPGA architecture.

V. A RAID ERROR CORRECTION CASE STUDY

The procedure for flat expressions is illustrated by the example of a RAID correction system. In RAID, system redundancy for error correction is kept using Reed–Solomon (RS) coding over Galois fields. Here we chose a field with 2^4 elements. We will concentrate on the implementation of the encoder/decoder parts and will talk of $n:m$ RS indicating an RS schema, where m checksum disks are used as redundancy for n disks of data.

Basically, the encoder part will take n streams to be stored in the data disks to generate the m streams to be stored in the redundant disks. This makes for a unique reconfiguration for a given $n:m$ schema.

On the other hand, the decoder part will take the streams stored on disks and return the original data streams. When all stored streams are available, the decoders simply return the data from the data disks. However, if one (or

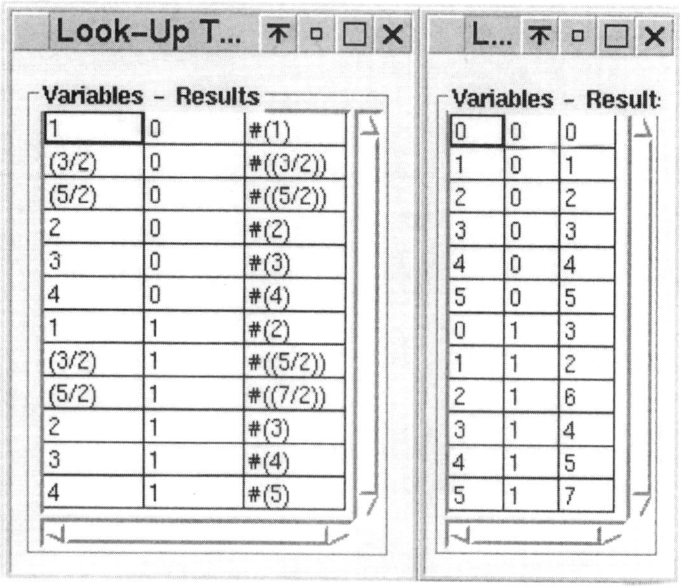

Look-Up T...

Variables - Results

1	0	#(1)
(3/2)	0	#((3/2))
(5/2)	0	#((5/2))
2	0	#(2)
3	0	#(3)
4	0	#(4)
1	1	#(2)
(3/2)	1	#((5/2))
(5/2)	1	#((7/2))
2	1	#(3)
3	1	#(4)
4	1	#(5)

L...

Variables - Result

0	0	0
1	0	1
2	0	2
3	0	3
4	0	4
5	0	5
0	1	3
1	1	2
2	1	6
3	1	4
4	1	5
5	1	7

Figure 6 Values and index lookup tables.

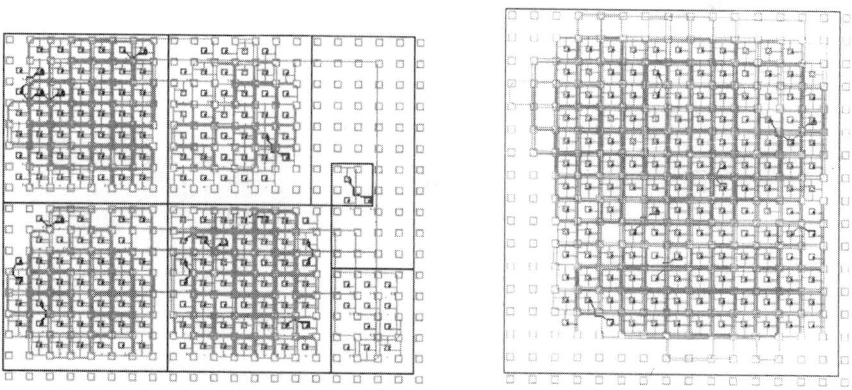

Figure 7 Hierarchical and flat circuit for the multiplier.

up to m) disks fails, the original data may need to be reconstructed from the checksums.

The use of FPGA for encoding/decoding seems appropriate mainly for two reasons:

There may be little or no performance loss due to error correction. The cost is paid only when a disk failure happens (transition from working disk to nonworking disk).

It will provide the system with added flexibility. The ability to mutate the circuits will allow the same hardware to be used for different failure schemas.

We have based our case study of the encoding/decoding in a word-by-word basis. This means that, as we are talking of Galois fields 2^4, we obtain circuits that take data from the streams in groups of 4 bits. As the operations for different words are independent, it is possible to replicate the circuits to work on multiple data words simultaneously (at the expense of logic space) to meet desired performance.

A. Reed–Solomon Coding

Reed–Solomon (RS) coding allows correction of up to m errors using m checksum words. For a system with n data words allowing error recovery for m failures, a total of $n + m$ words should be stored. The basic idea of RS coding is to build a system with $n + m$ rows and n columns. All rows are built to be independent. Recovery from up to m errors is possible as we can always take the available words and build a system that is solvable. Solving that system by any technique (like Gaussian elimination) will provide the original data words.

1. Encoding

At this point we need to build m additional independent equations. To achieve that, we will use the Vandermonde matrix and compute its associated independent terms that will serve as checksums. This can be done by performing the following operation (where $d_1 \ldots d_n$ are the data words to encode and $c_1 \ldots c_m$ its associated checksums):

$$\begin{bmatrix} 1 & 1 & 1 & \cdots & 1 \\ 1 & 2 & 3 & \cdots & n \\ \vdots & \vdots & \vdots & & \vdots \\ 1 & 2^{m-1} & 3^{m-1} & \cdots & n^{m-1} \end{bmatrix} \begin{bmatrix} d_1 \\ d_2 \\ \vdots \\ d_n \end{bmatrix} = \begin{bmatrix} c_1 \\ c_2 \\ \vdots \\ c_m \end{bmatrix} \tag{1}$$

2. Decoding

When retrieving data we know that the following equation must apply:

$$\begin{bmatrix} 1 & 0 & 0 & \cdots & 0 \\ 0 & 1 & 0 & \cdots & 0 \\ \vdots & \vdots & \vdots & & \vdots \\ 0 & 0 & 0 & \cdots & 1 \\ 1 & 1 & 1 & \cdots & 1 \\ 1 & 2 & 3 & \cdots & n \\ \vdots & \vdots & \vdots & & \vdots \\ 1 & 2_{m-1} & 3_{m-1} & \cdots & n_{m-1} \end{bmatrix} \begin{bmatrix} d_1 \\ d_2 \\ \vdots \\ d_n \end{bmatrix} = \begin{bmatrix} d_1 \\ d_2 \\ \vdots \\ d_n \\ c_1 \\ c_2 \\ \vdots \\ c_m \end{bmatrix} \tag{2}$$

So in case of an error on a word of data, we can compute its value by solving a system involving n rows of the equation. This will be possible as long as we have n valid values in the independent term vector; that is, there are less than m errors.

3. Arithmetic over Galois fields

This is used as the algebra needed to solve the system, as it is closed over a field of finite size. For a more detailed description of RS coding using arithmetic over Galois fields (GF), the reader may refer to Plank's tutorial [13], on which we have based our work. Another interesting bibliography comes from C. Paar et al., for example [14] provides information on GF operator complexity and implementation on FPGAs.

B. Encoder/Decoder Specifications

Our objective is to obtain configurations for the encoding/decoding using RS over Galois fields. We are targeting a system that has a reconfigurable part to do both encoding and decoding. We will need a configuration for the encoder and several configurations for decoding, one for each working condition (set of words missing). This can be applicable to RAID systems, where the working condition (disk failure) can be considered rare, and the cost of reconfiguration in such a case will have little impact (see Fig. 8).

The specifications have been developed in three steps:

Reed–Solomon specification. The specification generates the equation 2. This is specified as a Smalltalk class that has methods for encoding/decoding data. The class was built in order to fix the number of data and checksum words at instance creation, so it could be used for any encoding size. The specification was tested by exhaustively performing error correction using conventional arithmetic.

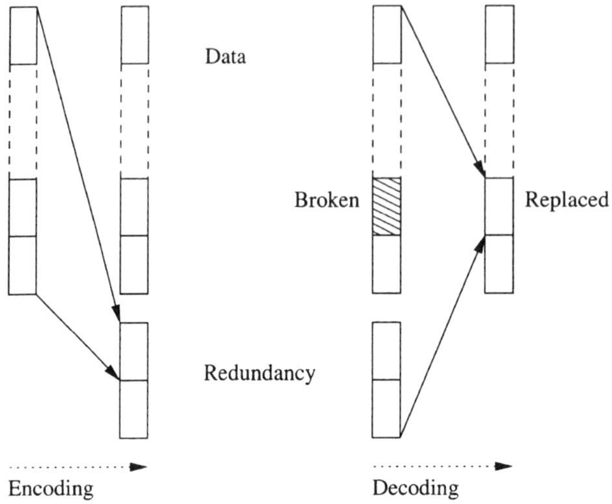

Figure 8 Encoding n data slices to $n + m$ disks with one redundant encoder shown. Decoding from $n + m$ disks to n data slices with one decoder for a broken disk shown.

Development of Galois field arithmetic class. A Galois field 2^4 for Smalltalk has been developed. The operations implemented are those needed in our problem, following the guidelines shown in [13]. Using Smalltalk's powerful polymorphic nature, we can apply the Reed–Solomon class using the new arithmetic. We performed an exhaustive test of the RS coding using the Galois field arithmetic in order to test correctness.

Extraction of error correction expressions. Building an arithmetic like class that records the operations performed, we can build the expressions for the encoding of checksums, as well as for the decoders in a given working condition. Those expressions can be packed into method code that will be compiled to build the configurations.

As a note, the above specifications took a few hours to be done, with no initial experience on Reed–Solomon coding.

C. Expression Compilation

From the specification, we can take the expressions for the encoders/decoders and compile them to logic. The steps below show the most important effect of the implementation as handled in our framework.

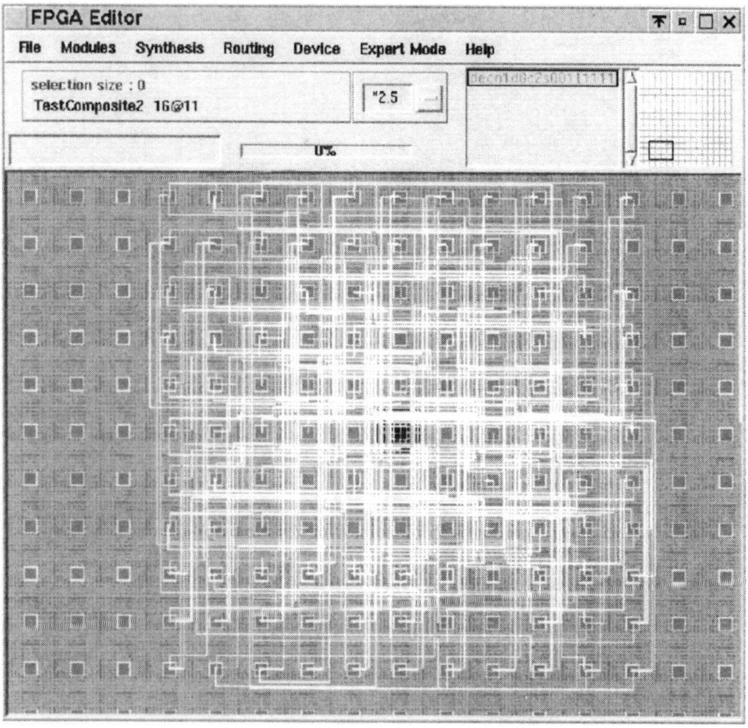

Figure 9 A decoder for an 8-data and two redundancy disks problem placed and routed on a LUT-2 architecture.

Type inference. After building the DAG, all edges are typed. In this case, the inputs are data and checksum words, all of type GF16.

Constant folding. Due to the generated nature of our expressions, there are plenty of operations over constants. Those are all removed in this steps. Also operations wielding a constant result (like multiplication by 0) are removed.

Dead code removal. As a result of removing multiplication by 0 operations, and due to the nature of automatically generated expressions, it is possible that there are expressions whose result is never used, so they are removed.

Code factorization. That is, common subexpression elimination.

Operator LUTification. This step is the first one toward architecture binding. It transforms the symbolic operations into look-up tables suitable for logic synthesis.

No-op removal. Unary operators whose output is equivalent to its input are removed.

Operator fusion. Unary operators are removed by fusing them with its producer/consumer operators. We assume that this will provide a better implementation.

Circuit production. Several circuits have been produced to collect practical information of the results.

D. Encoders/Decoders Statistics

Tables 1 and 2 display, respectively, statistics for the 3 necessary encoders and all the possible decoders for disk failures. The tables have columns showing the decrease in the number of GF16 operators, average number of inputs per operator, and the critical path in the network of operators as a result of each compiler operation. The meaning of the rows is the observed value after each optimization operation as described in subsection C.

Correctness has been checked at the logic level by selecting random random inputs and verifying the output *after logic synthesis* using an SIS *simulate* command.

E. Specific 8:2 Case with Circuit Generation

This time, the case of a RAID system with eight data disks and two redundant disks is considered. These circuits were optimized and mapped to two different architectures having 2-LUT (Figure 9) and 4-LUT cells. Table 3 shows the compared characteristics of the encoder and decoder on these architectures. Each has a routing channel of size 8 inside the cell patterns, providing a first-run success. Some parameters are extracted that can be used at a higher level, as an example for system management of the reconfigurable logic resources, or for making choices in the compiler generation code strategy. Notice that at

Table 1 Statistics for Encoders-RS4:3

Compiler operation	Operators	Average input	Critical path
Type inference	12	2	6
Constant folding	8	2	5
Dead-code removal	8	2	5
Code factorization	8	2	5
Operator to LUT	8	1.375	5
No-op removal	5	1.67	3.67
Operator fusion	3	2	3

Table 2 Statistics for Decoders-RS4:3

Compiler operation	Operators	Average input	Critical path
Type inference	85.08	2	11.24
Constant folding	11.68	2	7.65
Dead-code removal	11.68	2	7.65
Code factorization	10.41	2	7.65
Operator to LUTs	10.41	1.43	7.65
No-op removal	7.42	1.65	5.75
Operator fusion	4.5	2	3.625

the end of optimization on LUT, it is easy to generate processor code and table contents equivalent to the network of reconfigurable logic cells.

The table has two parts for post-assembly optimized logic and simple structured assembly as it is used for floor planning. The presented characteristics are:

1. Total circuit area in number of cell patterns
2. Gates used in this area
3. Number of inputs for the circuit
4. Effective number of cells used to implement logic
5. The average of used inputs in module cells including the border, and
6. The same measure for cells in (4).

Table 3 Results from Place and Route on Two Architectures

	Encoder		Decoder	
	LUT 2	LUT 4	LUT 2	LUT 4
Area (1)	90	56	121	72
Cells used (2)	85	53	119	71
Input cells (3)	32	32	40	40
Internal cells (4)	53	21	79	31
Input average (5)	1.62	2.04	1.67	2.23
Gates input average (6)	2.0	3.62	2.0	3.81
Routing cost (7)	1095	640	1839	1049
Critical path (8)	18	14	19	15
CPU time (9)	43.14	20.34	98.70	34.89
Max struct. area (10)	128	88	208	176
Cells used (11)	109	85	181	170
Internal cells (12)	53	29	79	58

Notice that (3) + (4) = (2), with (4) being low. The circuit is I/O dominated. Gates used in (3) disappear when the module is connected to other architecture elements.

Routing cost (7) is an estimation on the number of resources allocated for connections. Critical path (8) is the maximum number of cells and other resources allocated in the circuit between an input and an output, with unitary costs. CPU time (9) provides an idea of the delay to place and route the circuit on a PC/750Mhz with the Visualworks environment running on Linux. Figure 9 show a view of a decoder as generated by the tools.

The maximum area occupied by the assembly of elementary modules is (10) without post-assembly optimization, and without the use of the floor planner. The maximum number of cells used in this area is (11), and (12) is the number of cells used to implement logic in the area. Item (12) is similar to (4). A good measure of the post-assembly optimization is the respective 46% and 27% logic decreases in the cases of the decoder and encoder. The use of the floor planner will bring (10) and (11) closer to (1) and (2).

VI. CONCLUSION

Madeo tools for reconfigurable architecture modeling are operational with practical implementations on commercial FPGAs and prospective FPGA prototype architectures. This demonstrates the feasibility of a FPGA hardware/software interface standardization in a way similar to microprocessors and HLL compilers.

The described compiler is still a work in progress. It is possible to produce an optimized hierarchical logic description suitable for technology mapping, place, and route. Dependencies in fan-out to fan-in subgraph, rejecting unused values from the branches of conditional statements, and recursive specifications are handled. The main effect of this compiler is to achieve optimizations mostly at a high level, removing a considerable load on the logic mapping algorithms. The strength in optimization comes from the fine knowledge on values being processed, that allows to simplify computations either at a high level or logic level. The underlying execution model is understandable for the programmer who has direct feedback for the algorithms.

Related to productivity in developments, our method also allows the possibility to create specific logic, based on concise behavioral specifications that are reusable in a variety of situations on different kinds of data. The compiler is able to allocate medium-grain resources such as memories or arithmetic operators, based on the architectural model and types propagated inside the computation graph.

Finally, we find the object-oriented approach very promising either for architecture management or high-level synthesis. This encourages more research in this direction for higher scope work in the context of systems on chips.

ACKNOWLEDGMENTS

This work was supported by grants from the MEFI/STSI and STMicroelectronics. The authors thank O. Colavin, T. Ben-Ismail, and V. George for all they shared and many thanks to all the students involved in the various aspects of development.

REFERENCES

1. Debaere, E., Campenhout, J. M. V. (1990). *Interpretation and Instruction Path Coprocessing*. MIT Press.
2. Lagadec, L., Pottier, B. (November 2000). Object oriented meta-tools for reconfigurable architectures. *SPIE, Reconfigurable Technology II*. Vol. 4212.
3. Sentovich, E. M., et al. (1992). Sis A system for sequential circuit synthesis. Technical report, EECS, Berkeley.
4. Lemarchand, L. (1999). Parallel performance directed technology mapping for fpga. *SSMSD*. Tucson: IEEE.
5. Lin, B., Whitcomb, S., Newton, A. (1991). Symbolic don't care and equivalence in high level synthesis. In: IFIP, ed. *Logic and Architecture Synthesis*. Elsevier.
6. Ballard, M. B., Wirfs-Brock, A. (Nov. 1986). Quicktalk: a smalltalk-80 dialect for defining primitive methods. *OOPSLA '86 proc.* 21,(11).
7. Llopis, J. -L., Pottier, B. (1996). Smalltalk blocks revisited, a logic generator for fpgas. *FCCM'96*.
8. Fabregat, G., Leon, G., Le Berre, O., Pottier, B. (Oct. 1999). Embedded system modeling and synthesis in oo environments. a smart-sensor case study. *CASES'99*.
9. Weaver, N., Chu, M., Sulimma, K., Dehon, A. (Apr. 1998). Object oriented circuit-generators in java. *FCCM*.
10. Peymandoust, A., De Micheli, G. (2001). Using symbolic algebra in algorithmic level dsp synthesis. *DAC*.
11. Bertin, Touati. Pam programming environments: practice and experience. *FCCM* 133–138.
12. Patterson, D.A., Hennessy, J. (1998). *Computer Organisation and Design, the hardware/software interface*. Morgan Kaufmann.
13. Plank, J. (1999). A tutorial on reed-solomon coding for fault-tolerance in raid-like systems. Technical report, DCS, University of Tennessee.
14. Paar, C., Rosner, M. (1997). Comparison of arithmetic architectures for reed-solomon decoders in reconfigurable hardware. *FCCM'97*.

3

Highly Efficient Scalable Parallel-Pipelined Architectures for Discrete Wavelet Transforms

David Guevorkian, Petri Liuha, Aki Launiainen, and Ville Lappalainen
Nokia Research Center, Tampere, Finland

I. INTRODUCTION

Discrete wavelet transforms (DWTs) have been extensively developed as an efficient multiresolution analysis tool during the past decade. They provide an efficient technique for signal/image decomposition into different subbands of well-defined time-frequency characteristics [1–5]. Wavelets have been studied and successfully applied to a wide range of applications such as different branches of image and video processing, numerical analysis, biomedicine, signal processing techniques, speech compression/decompression, etc. DWT-based compression methods have become the basis of international standards such as JPEG 2000.

Since many applications need real-time computation of DWT, a number of ASIC architectures have already been proposed (see [6–25]) for hardware implementation of DWTs. Most of the early architectures exploit the recursive pyramid algorithm (RPA) [26], based on the tree-structured filter bank representation of DWTs (see Fig. 1a). In this representation the input signal is processed by several (J) levels of decomposition (octaves), where the input at every stage is processed by a low-pass and by a high-pass filter, outputs of which are then downsampled by a factor of two. The length of the processed signal is twice reduced from level to level. Based on this representation of DWTs several low-hardware complexity devices have been developed that

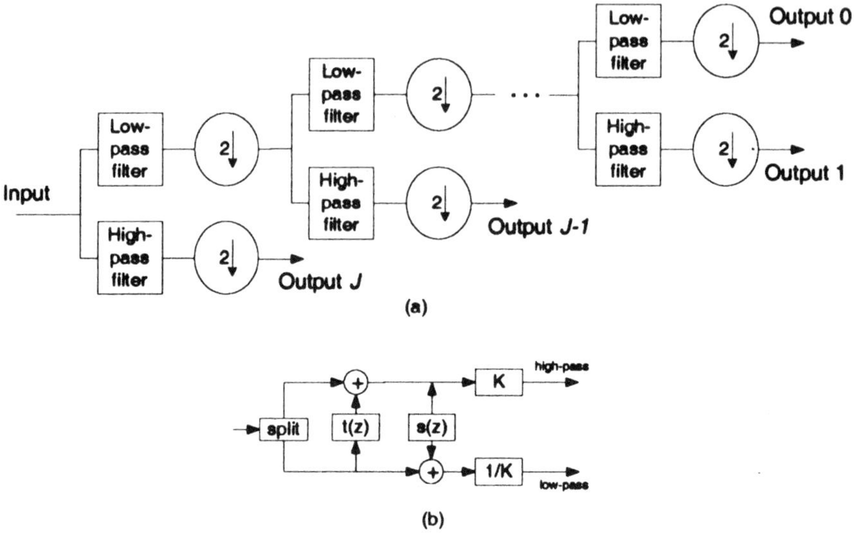

Figure 1 (a) Tree-structured flowgraph representation of a 1-D DWT; (b) lifting step.

require at least $2N$ clock cycles (ccs) to compute a DWT of a sequence having N samples (e.g., the devices proposed in [12–14], the architecture A2 in [15], etc.). Also a large number of devices, having a period of approximately N cc's, have been designed (e.g., the three architectures in [14] when they are provided with a doubled hardware, the architecture A1 in [15], the architectures in [16–18], the parallel filter in [19], etc.). In order to increase the throughput, pipelining has been employed to implement these structures where every DWT octave is implemented in a pipeline stage (pipelining at the octave level). However, despite the fact that the amount of computations reduces twice from one stage to the next, these pipelined architectures use the same number of processing elements (PEs) for every pipeline stage [12], [23–25]). Balancing of the pipeline stages is achieved by making the clocking fequency twice as small from stage to stage (see, e.g., [23–25]). This is clearly a pure hardware utilization of PEs.

Recently, a number of DWT architectures [6–11] have been developed based on another representation of wavelets, on the lifting scheme where the filtering and downsampling procedure is replaced by so called lifting steps as shown in Fig. 1b. This is a more efficient approach and allows reducing the memory sizes and accesses (which are crucial especially, in the 2–D case) [6–10], as well as the processing time [11]. However, to the best of our knowledge no lifting-based DWT architectures that exploit pipelining of DWT octaves have been proposed. The reason is, perhaps, the same as for tree-structured filter bank representation-based architectures. That is, as the length of the signal is twice reduced from one stage to the next, it is difficult to design a pipelined architecture with well-balanced pipeline stages. Low-balancing between the pipeline stages leads to under-utilization of the PEs.

As a consequence of under-utilization of PEs, the typical efficiency of conventional pipelined architectures strongly decreases with the number of decomposition levels. Approximately 100% of efficiency is achieved only in conventional architectures that are either nonpipelined or employ only a restricted (two-stage) pipelining (e.g. those in [17,18]). These architectures, in addition, suffer from extensive memory (chip area) or control requirements. The highest throughput achieved in known architectures is $N/2$ clock cycles per N-point DWT (see [11]). Similar performance is achieved in highly (about 100%) efficient architectures developed in [27–28] by including approximately twice the lower number of PEs from stage to stage.

In [29–30], *flowgraph representation* of DWTs (see examples in Figs. 2 and 3) has been suggested as a useful tool in designing parallel/pipelined DWT architectures. In particular, this representation fully reveals parallelism

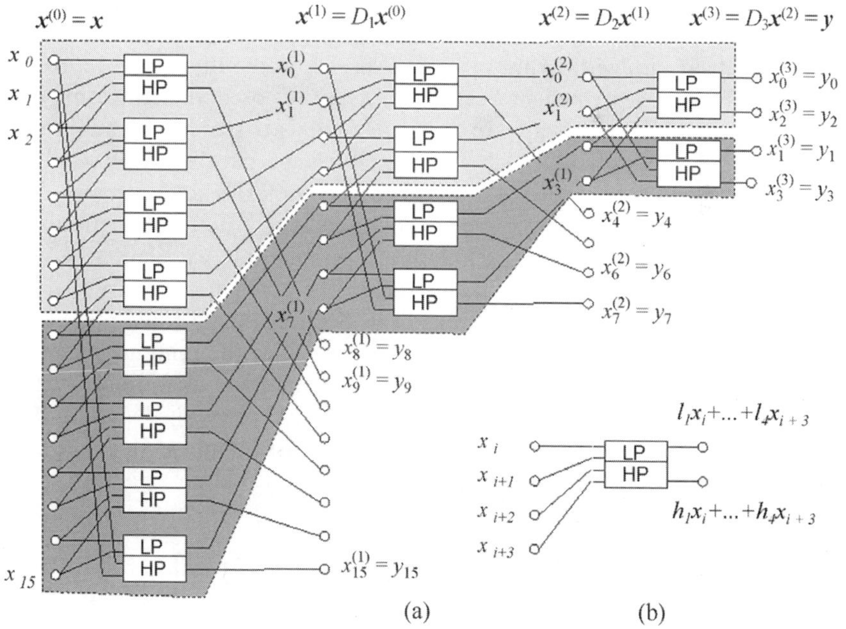

Figure 2 Flowgraph representation of a DWT ($N=16$, $L=4$, $J=3$).

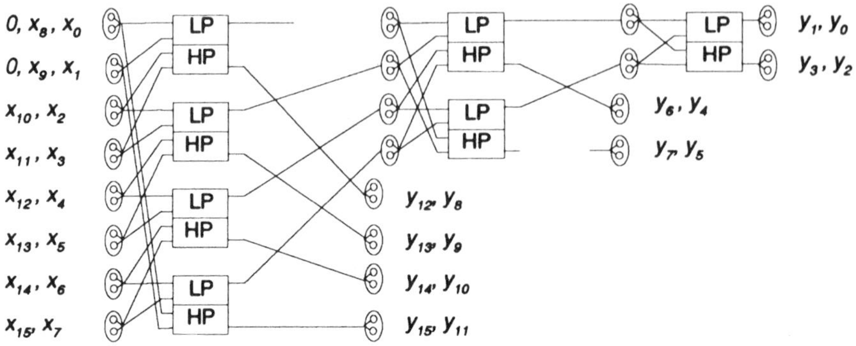

Figure 3 Compact flowgraph representation of a 1-D DWT ($L = 4$, $J = 3$).

inherent to every octave as well as it clearly demonstrates data transfers within and between octaves. This allows combining pipelining and parallelism to achieve a higher cost-efficient performance. This means implementing octaves in a pipelined mode where pipeline stages are parallelized at varying stage-to-stage levels. Incorporating varying level parallelism within pipeline stages allows the design of parallel pipelined devices with perfectly balanced pipeline stages.

In this work, several DWT architectures operating at approximately 100% hardware utilization are proposed, based on the flowgraph representation of DWTs described in Section II. The proposed structures may be implemented in different ways. In particular, they are scalable meaning that they can be implemented with varying levels of parallelism, giving an opportunity to trade-off between hardware complexity and performance. Several possible realizations of the proposed general structures are discussed in Section III. The resulting architectures demonstrate excellent time and moderate area performance as compared to the conventional DWT architectures, as follows from the discussion in Section IV. Throughput of the architectures may vary between $NL/2^J$ time units (at the minimum level of parallelism) up to even one time unit (at the theoretical maximum level of parallelism) per N-point DWT with J octaves and filters of the length L. The proposed architectures are regular and modular, easily controlled, and free of feedback or a switch. They can be implemented as semisystolic arrays.

II. FLOWGRAPH REPRESENTATION AND PARALLEL ALGORITHMS FOR DWTs

There are several alternative definitions and representations of DWTs such as tree-structured filter bank, lattice structure, lifting scheme, or matrix definitions [1–11]. In this section, we use the matrix definition of DWTs to arrive at their flowgraph representation, which has been shown to be very efficient in designing efficient parallel/pipelined DWT architectures [29–30]. The basic algorithm for the proposed structures is also described in this section.

Using the matrix definition, a discrete wavelet transform is a linear transform $y = H \cdot x$, where $x = [x_0,..., x_{N-1}]^T$ and $y = [y_0,..., y_{N-1}]^T$ are the input and the output vectors of length $N = 2^m$, respectively. H is the DWT matrix of order N x N, which is formed as the product of sparse matrices:

$$H = H^{(J)} \cdot \ldots \cdot H^{(1)}, \qquad 1 \leq J \leq m; \qquad H^{(j)} = \begin{pmatrix} D_j & 0 \\ 0 & I_{2^m - 2^{m-j+1}} \end{pmatrix}$$

$$j = 1, \ldots, J \tag{1}$$

where I_k is the identity $(k \times k)$ matrix $(k = 2^m - 2^{m-j+1})$. D_j is the analysis $(2^{m-j+1}$ x $2^{m-j+1})$ matrix at stage j having the following structure:

$$D_j = \begin{pmatrix} l_1 & l_2 & \ldots & l_L & 0 & 0 & \ldots & 0 \\ 0 & 0 & l_1 & l_2 & \ldots & l_L & \ldots & 0 \\ & & & \ddots & & & & \\ l_3 & \ldots & l_L & 0 & 0 & \ldots & l_1 & l_2 \\ h_1 & h_2 & \ldots & h_L & 0 & 0 & \ldots & 0 \\ 0 & 0 & h_1 & h_2 & \ldots & h_L & \ldots & 0 \\ & & & \ddots & & & & \\ h_3 & \ldots & h_L & 0 & 0 & \ldots & h_1 & h_2 \end{pmatrix}$$

$$= P_j \begin{pmatrix} l_1 & l_2 & \ldots & l_L & 0 & 0 & \cdots & 0 \\ h_1 & h_2 & \ldots & h_L & 0 & 0 & \ldots & 0 \\ 0 & 0 & l_1 & l_2 & \ldots & l_L & \ldots & 0 \\ 0 & 0 & h_1 & h_2 & \ldots & h_L & \ldots & 0 \\ & & & \ddots & & & & \\ l_3 & \ldots & l_L & 0 & 0 & \ldots & l_1 & l_2 \\ h_3 & \ldots & h_L & 0 & 0 & \ldots & h_1 & h_2 \end{pmatrix} \tag{2}$$

where $LP = [l_1,..., l_L]$ and $HP = [h_1,..., h_L]$ are the vectors of coefficients of the low-pass and high-pass filters, respectively, L is the length of the filters*, and P_j is the matrix of the perfect unshuffle operator of size $(2^{m-j+1} \times 2^{m-j+1})$.

Adopting the representation from Eqs. (1–2), the DWT is computed in J stages (also called decomposition levels or octaves), where the jth stage,

* For clarity we assume both filters to have the same length, which is an even number. The results are easily expanded to the general case of arbitrary filter lengths.

$j = 1,\ldots, J$, constitutes a multiplication of a sparse matrix $H^{(j)}$ by a current vector of scratch variables where the first such vector being the input vector x. The corresponding algorithm can be written as the following pseudocode where $x_{LP}^{(j)} = [x_{LP}^{(j)}(0),\ldots, x_{LP}^{(j)}(2^{m-j}-1)]^T$, and $x_{HP}^{(j)} = [x_{HP}^{(j)}(0),\ldots, x_{HP}^{(j)}(2^{m-j}-1)]^T$, j $1)]^T$, $j = 1,\ldots, J$, are $(2^{m-j} \times 1)$ vectors of scratch variables, and the concatenation of column vectors $x_1,\ldots, x_k$ is denoted as $[(x_1)^T,\ldots, (x_k)^T]^T$.

Algorithm 1
1. Set $x_{LP}^{(0)} = \left[x_{LP}^{(0)}(0),\ldots,x_{LP}^{(0)}(2^m-1)\right]^T = x$;
2a. For $j = 1,\ldots, J$ compute

$$x_{LP}^{(j)} = \left[x_{LP}^{(j)}(0),\ldots,x_{LP}^{(j)}(2^{m-j}-1)\right]^T \text{ and}$$

$$x_{HP}^{(j)} = \left[x_{HP}^{(j)}(0),\ldots,x_{HP}^{(j)}(2^{m-j}-1)\right]^T,$$

where

$$\left[\left(x_{LP}^{(j)}\right)^T, \left(x_{HP}^{(j)}\right)^T\right]^T = D_j \cdot x_{LP}^{(j-1)} \tag{3}$$

or, equivalently,
2b. For $i = 0,\ldots, 2^{m-j} - 1$,
Form the vector /* subvector of length L of the vector $x_{LP}^{(j)}$ */

$$\tilde{x} = \left[x_{LP}^{(j-1)}(2i), x_{LP}^{(j-1)}(2i+1),\ldots,x_{LP}^{(j-1)}\left((2i+L-1) \bmod 2^{m-j+1}\right)\right]^T$$

Compute

$$x_{LP}^{(j)}(i) = LP \cdot \tilde{x}; \quad x_{HP}^{(j)}(i) = HP \cdot \tilde{x}$$

3. Form the output $y = [x_{LP}^{(J)}, x_{HP}^{(J)}, x_{HP}^{(J-1)},\ldots, x_{HP}^{(2)}, x_{HP}^{(1)}]^T$.

Computation of Algorithm 1 with the matrices Dj of Eq. (2) can be clearly demonstrated using a flowgraph representation. An example for the case $N = 2^3 = 8$, $L = 4$, $J = 3$ is shown in Fig. 2. The flowgraph consists of J stages, the j-th stage, $j = 1,\ldots, J$, having 2^{m-j} nodes (depicted as boxes on Fig. 2). Each node represents a basic DWT operation (see Fig. 2b). The ith node, $i = 0,\ldots, 2^{m-j} - 1$, of the stage $j = 1,\ldots, J$ has incoming edges from L circularly consecutive nodes $2i$, $2i + 1$, $(2i + 2) \bmod 2^{m-j+1},\ldots, (2i + L - 1) \bmod 2^{m-j+1}$ of the preceding stage or (for the nodes of the first stage) from inputs. Every node has two outgoing edges. An upper (lower) outgoing edge represents the value of the inner product of the vector of low-pass (high-pass) filter coefficients with the vector of the values of incoming edges. Outgoing values of a stage are permuted according to the perfect unshuffle operator so that all the low-pass components are collected in the first half and the high-pass components are collected at the second half of the permuted vector. Low-pass components are then forming the input to the following stage or (for the nodes of the last stage) represent output values. High-pass components represent output values at a given resolution.

Essentially, the flowgraph representation gives an alternative definition of discrete wavelet transforms which is rather illustrative and easy-to-understand. It has several advantages, at least from an implementation point of view, as compared to the conventional DWT representations such as the tree-structured filter bank, lifting scheme, or lattice structure representations [30].

The flowgraph representation, however, of DWTs as presented so far, has the inconvenience of being very large for bigger values of N. This inconvenience can be overcome based on the following observation. Assuming $J < \log_2 N$ (in most of applications $J << \log_2 N$), one can see that the DWT flowgraph consists of $N/2^J$ similar patterns (see the two shadowed regions on Fig. 2). Every pattern can be considered as a 2^J-point DWT with a specific strategy of forming the input signals to its every octave. Merging the 2^{m-J} patterns in one, we can now obtain a *compact* (*or core*) *flowgraph* representation of DWT. An example of a DWT compact flowgraph representation for the case $J = 3$, $L = 4$ is shown in Fig. 3. The compact DWT flowgraph has 2^{J-j} nodes at its j-th, stage, $j = 1,..., J$, where now a set of 2^{m-J} temporarily distributed values are assigned to every node. Also, every outgoing edge corresponding to a high-pass filtering result of a node or low-pass filtering result of the node of the last stage represents a set of 2^{m-J} output values.

Note that the structure of the compact DWT flowgraph does not depend on the length of the DWT but only on the number of decomposition levels and filter length. The DWT length is reflected only in the number of values represented by every node.

To illustrate the computation represented by the compact flowgraph let us adopt following notations. Let $\hat{\mathrm{D}}j$ be a matrix consisting of the first 2^{J-j+1} rows and the first $2^{J-j+1} + L - 2$ columns of $\mathrm{D}j$. Let us also conventionally divide the vector $x_{LP}^{(j-1)}$ of Eq. (3) into subvectors $x^{(j-1,s)} = x_{LP}^{(j-1)}(s \cdot 2^{J-j+1}: (s+1)2^{J-j+1} - 1)$, $s = 0,..., \cdot 2^{m-J} - 1$, where the notation $x(a: b)$ stands for the subvector of x consisting of the ath to bth components of x. Then the input of the j-th, $j = 1,..., J$, octave within the s-th pattern of the DWT flowgraph is the subvector $\hat{\mathrm{x}}^{(j-1,s)}(0: 2^{J-j+1} + \mathrm{L} - 3)$ of the vector

$$\hat{x}^{(j-1,s)} = \left[\left(x_{LP}^{(j-1,s \bmod 2^{m-J})} \right)^T, \ldots, \left(x_{LP}^{(j-1,(s+Q_j-1) \bmod 2^{m-J})} \right)^T \right]^T \quad (4)$$

being the concatenation of the vector $x_{LP}^{(j-1,s)}$ with the circularly next $Q_j - 1$ vectors where $Q_j = \lceil \cdot (L-2)/2^{J-j+1} \rceil$. With these notations, the computation of the compact flowgraph can be described with the following pseudocode.

Algorithm 2
1. For $s = 0,\ldots, 2^{m-J} - 1$ set $x_{LP}^{(0,s)} = x(s\cdot 2^J: (s+1)\cdot 2^J - 1)$;
2. For $j = 1,\ldots, J$
For $s = 0,\ldots, 2^{m-J} - 1$
2.1 Set $\hat{x}^{(j-1,s)}$ according to (4)
2.2 Compute

$$\left[\left(x_{LP}^{(j,s)}\right)^T, \left(x_{HP}^{(j,s)}\right)^T\right]^T = \hat{D}_j \cdot \hat{x}^{(j-1,s)}\left(0 : 2^{J-j+1} + L - 3\right)$$

3. Form the output $y = \left[\left(x_{LP}^{(J,0)}\right)^T, \ldots, \left(x_{LP}^{(J,2^{m-J}-1)}\right)^T, \left(x_{HP}^{(J,0)}\right)^T, \ldots, \left(x_{HP}^{(J,2^{m-J}-1)}\right)^T, \ldots, \left(x_{HP}^{(1,0)}\right)^T, \ldots, \left(x_{HP}^{(1,2^{m-J}-1)}\right)^T\right]^T$

Implementing the cycle for s in parallel one can easily arrive to a parallel DWT realization. By exchanging the nesting order of cycles for j and s and implementing the (nested) cycle for j in parallel it is possible to arrive to a pipelined DWT realization. Both poorly parallel and poorly pipelined realizations would be inefficient since the number of operations is halved from one octave to the next one. However, combining the two methods we arrive at a very efficient parallel-pipelined or partially parallel-pipelined realization. The following pseudocode presents such parallel-pipelined DWT realization where we denote

$$s*(j) = \sum_{n=1}^{j} Q_n \qquad (5)$$

Algorithm 3
1. For $s = 0,\ldots, 2^{m-J} - 1$ set $x_{LP}^{(0,s)} = x(s\cdot 2^J : (s+1)\cdot 2^J - 1)$;
2. For $s = s*(1),\ldots, 2^{m-J} + s*(J) - 1$, For $j = J_1,\ldots, J_2$ do in parallel
2.1 Set $\hat{x}^{(j-1,s-s*(j))}$ according to (4)
2.2 Compute

$$\left[\left(x_{LP}^{(j,s-s*(j))}\right)^T, \left(x_{HP}^{(j,s-s*(j))}\right)^T\right]^T = \hat{D}_j \cdot \hat{x}^{(j-1,s-s*(j))}\left(0 : 2^{J-j+1} + L - 3\right)$$

3. Form the output (See Step 3 of Algorithm 2.)

Note that computation of Algorithm 3 takes place during the steps $s = s*(j),\ldots, 2^{m-J} + s*(j) - 1$. At step $s = s*(1),\ldots,s*(2) - 1$ computations of only the first octave are implemented, at step $s = s*(2),\ldots,s*(3) - 1$ operations of the first two octaves are implemented, etc. In general, at the step

$s = s*(1),...,2^{m-J} + s*(J) - 1$ operations of the octaves $j = J_1,...,J_2$ are implemented where $J_1 = \min\{j \text{ such that } s*(j) \geq s < s*(j) + 2^{m-J}\}$ and $J_1 = \max\{j \text{ such that } s*(j) \leq s < s*(j) + 2^{m-J}\}$.

III. THE PROPOSED DWT ARCHITECTURES

In this section, we present general structures of two types of DWT architectures, referred to as *Type 1* and *Type 2 core DWT architectures*, as well as two other DWT architectures extended from either core DWT architecture and referred to as multicore DWT architecture and variable resolution DWT architecture, respectively. Both types of the core DWT architectures implement arbitrary discrete wavelet transform with J octaves with low-pass and high-pass filters having a length L not exceeding a given number $L_{\max}$.

The general structure representing both types of core DWT architectures is presented on Fig. 4 where dashed lines depict connections, which are present in Type 2 architectures but are absent in Type 1 architectures. In both cases the architecture consists of a data input block and J pipeline stages with each stage containing a data routing block and a block of processor elements (PEs) where the data input block implements Step 1 of Algorithm 3, data routing blocks are responsible for Step 2.1, and blocks of PEs are for computations of Step 2.2. The two types mainly differ by the possibility of data exchange between PEs at the same pipeline stage, which are possible in Type 2 but not in Type 1 architectures.

The data input block of the core DWT architectures of both types may be realized as word-serial as well as word-parallel. In the former case, the data input block consists of a single (word-serial) input port that is connected to a length-2^J shift register (dashed lined box in Fig. 4) having a word-parallel output from every cell. In the latter case, the data input block simply consists of 2^J parallel input ports. In both cases the data input block has 2^J parallel outputs connected to the 2^J inputs of the data routing block of the first pipeline stage.

A. Type 1 Core DWT Architecture

The basic operation of Algorithm 3 (Step 2.2) is equivalent to 2^{J-j} pairs of vector–vector inner products:

$$
\begin{aligned}
&x^{(j,s-s*(j))}(i) = LP \cdot \hat{x}^{(j-1,(s-s*(j))+)}(2i : 2i + L - 1) \\
&x^{(j,s-s*(j))}(i + 2^{J-j}) = HP \cdot \hat{x}^{(j-1,(J,s-s*(j))+)}(2i : 2i + L - 1) \\
&i = 0, \ldots, 2^{J-j} - 1
\end{aligned}
$$

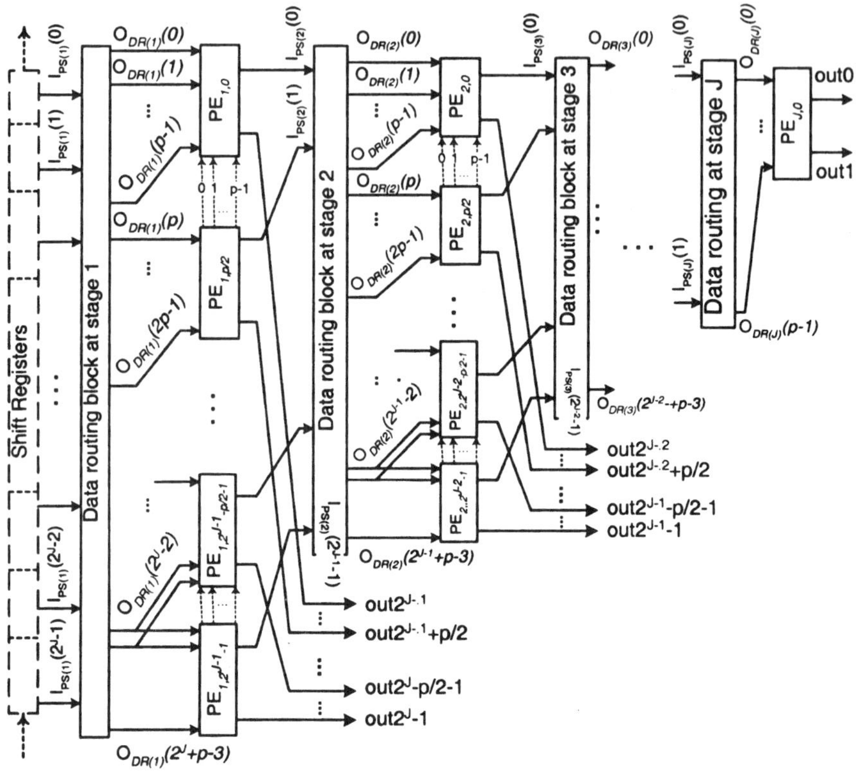

Figure 4 The general structure of Type 1 and Type 2 core DWT architectures.

All the inner products may be implemented in parallel. On the other hand, every vector–vector inner product of length L can be obviously decomposed into an accumulative sequence of $L_p = \lceil L/p \rceil$ inner products of length p ($p \leq L$). Algorithm 3 may be modified as follows.

Algorithm 3.1
1. For $s = 0, \ldots, 2^{m-J} - 1$ set $x_{LP}^{(0,s)} = x(s \cdot 2^J : (s+1) \cdot 2^J - 1)$;
2. For $s = s*(1), \ldots, 2^{m-J} + s*(J) - 1$, For $j = J_1, \ldots, J_2$ do in parallel
2.1. Set $\hat{x}^{(j-1,s-s*(j))}$ according to (4)
2.2. For $i = 0, \ldots, 2^{J-j} - 1$ do in parallel
Set $S_{LP}(i) = 0$, $S_{HP}(i) = 0$;
For $n = 0, \ldots, L_p - 1$ do sequentially

$$S_{LP}(i) = S_{LP}(i) + \sum_{k=0}^{p-1} l_{np+k} \hat{x}^{(j-1,s-s*(j))}(2i + np + k) \tag{6}$$

$$S_{HP}(i) = S_{HP}(i) + \sum_{k=0}^{p-1} h_{np+k}\hat{x}^{(j-1,s-s*(j))}(2i+np+k) \tag{7}$$

Set $x_{LP}^{(j,s-s*(j))}(i) = S_{LP}(i)$; $x_{HP}^{(j,s-s*(j))}(i) = S_{HP}(i)$

3. Form the output (see Step 3 of Algorithm 2).

The general structure of the Type 1 core DWT architecture is presented in Fig. 4 where the dashed lines should be ignored, that is, there are no connections between PEs of a single stage. The architecture consists of the above-described data input block and J pipeline stages. In general, the jth pipeline stage, $j = 1,\ldots, J$, of the Type 1 core DWT architecture consists of a data routing block having 2^{J-j+1} inputs $I_{PS(j)}(0),\ldots, I_{PS(j)}(2^{J-j+1} - 1)$ forming the input to the stage, and $2^{J-j+1} + p - 2$ outputs $O_{DRB(j)}(0),\ldots, O_{DRB(j)}(2^{J-j+1} + p - 3)$ connected to the inputs of 2^{J-j} PEs.

Every PE has p inputs and two outputs where $p \leq L_{\max}$ is a parameter describing the level of parallelism of every PE. Consecutive p outputs $O_{DRB(j)}(2i), O_{DRB(j)}(2i+1),\ldots, O_{DRB(j)}(2i+p-1)$ of the data routing block of the jth, $j = 1,\ldots, J$, stage are connected to the p inputs of the ith, $i = 0,\ldots, 2^{J-j} - 1$, PE ($PE_{j,i}$) of the same stage. First outputs of 2^{J-j} PEs of the jth pipeline stage, $j = 1,\ldots, J-1$, form the outputs $O_{PS(j)}(0),\ldots, O_{PS(j)}(2^{J-j} - 1)$ of that stage and are connected to the 2^{J-j} inputs $I_{PS(j+1)}(0),\ldots, I_{PS(j+1)}(2^{J-j} - 1)$ of the data routing block of the next, $(j+1)$st, stage. The first output of the (one) PE of the last, Jth, stage is the 0th output out(0) of the architecture. Second outputs of 2^{J-j} PEs of the jth pipeline stage, $j = 1,\ldots, J$, form the (2^{J-j})th to $(2^{J-j+1} - 1)$st outputs $out(2^{J-j}),\ldots, out(2^{J-j+1} - 1)$ of the architecture.

Let us now describe functionality of the blocks of Type 1 core DWT architecture. For convenience, we define time unit as the period for PEs to complete their one operation (i.e., the period between successive groups of p data to enter to the PE) and let us consider an operation step of the architecture to consist of L_p time units.

The data input block serially (or in parallel) accepts and in parallel outputs a group of components of the input vector at the rate of 2^J components per operation step. Thus, the vector $x_{LP}^{(0,s)}$ is formed on the outputs of the data input block at the step $s = 0,\ldots, 2^{m-J} - 1$.

The data routing block of stage $j = 1,\ldots, J$, is a circuitry which at the first time unit $n = 0$ of its every operation step accepts in parallel a vector of 2^{J-j+1} components, and then at every time unit $n = 0,\ldots, L_p - 1$ of that operation step it outputs in parallel a vector of $2^{J-j+1} + p - 2$ components $np, np+1,\ldots, (n+1)p + 2^{J-j+1} - 3$ of a vector being the concatenation (in chronological order) of the vectors accepted at previous $\hat{Q}_j - 1$ steps, where

$$\hat{Q}_j = \lceil (L_{\max} - 2)/2^{J-j+1} \rceil \quad j = 1,\ldots, J \tag{8}$$

The functionality of the PEs used in Type 1 core DWT architecture is to compute two inner products Eqs. (6) and (7) of the vector on its p inputs with two vectors of predetermined coefficients during every time unit and to accumulate the results of both inner products computed during one operation step. At the end of every operation step, the two accumulated results pass to the two outputs of the PE and new accumulation starts. Possible structures of PEs for Type 1 core DWT architectures are presented in Fig. 5 for the case of arbitrary $pp = 1$, $p = 2$, and $p = L_{\max}$, (Fig. 5a–d, respectively). These structures are for the "generic" DWT implementation independent of filter coefficients. (They can be easily optimized for specific filter coefficients.) In particular, correlation between the low-pass and the high-pass filter coefficients may be used in order to reduce hardware requirements. Except this, PEs implementing the lifting steps (see [4]), similar to those in [10], and [11] may be used in the proposed architectures.

It is easy to show that the architecture implements computations according to Algorithm 3.1 though with extra delay when $L < L_{\max}$. The extra delay is the consequence of the flexibility of the architecture for being capable of implementing DWTs with arbitray filter length $L \leq L_{\max}$ while Algorithm 3.1 presents computation of a DWT with a fixed filter length L. In fact, the architecture is designed for filter length $L_{\max}$ but also implements

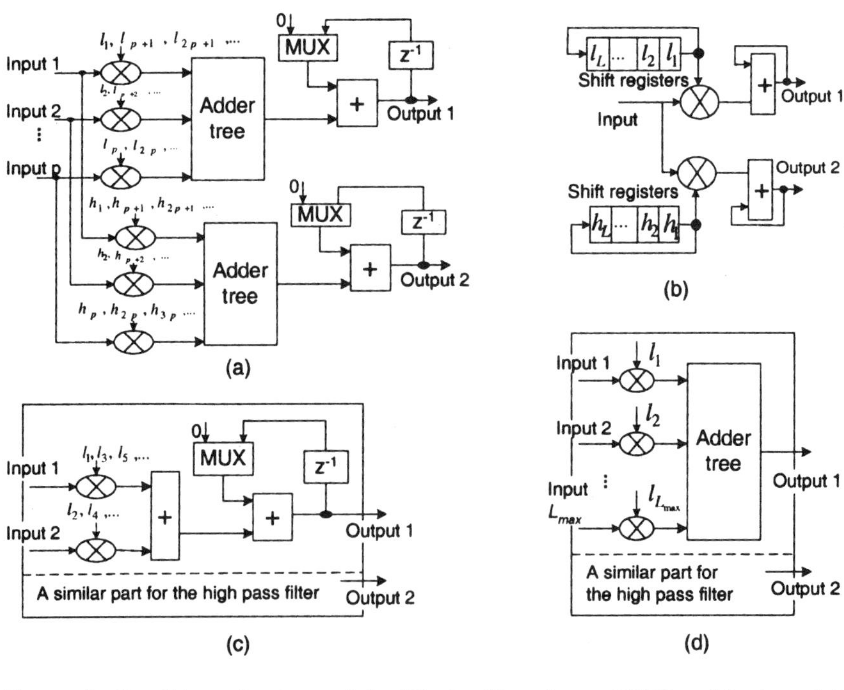

Figure 5 Possible realizations of the PEs for Type 1 core DWT architecture: (a) arbitrary p; (b) $p = 1$; (c) $p = 2$; and (d) $p = L_{max}$.

DWTs with shorter filters with a slightly increased time delay but without loosing the time period.

Denote

$$\hat{s}(0) = 0, \quad \hat{s}(j) = \sum_{n=1}^{j} \hat{Q}_n, \quad j = 1, \ldots, J \qquad (9)$$

The delay between input and output vectors is equal to

$$T_d(C1) = \left(2^{m-J} + \hat{s}(J)\right)\lceil L/p \rceil \qquad (10)$$

time units. The throughput or the time period (measured as the the intervals between time units when successive input vectors enter to the architecture) is equal to $T_p(C1)$ time units, where

$$T_p(C1) = 2^{m-J} \lceil L/p \rceil \qquad (11)$$

From Eqs. (10) and (11) we obtain that approximately 100% efficiency (hardware utilization* is achieved for the architecture both with respect to time delay and, moreover, time period complexities. A close efficiency is reached only in a few pipelined DWT designs (see [17, 27, 28]), but most of the known pipelined DWT architectures reach much less than 100% average efficiency. It should also be noted that a time period of at least $O(N)$ time units is required by known DWT architectures.

The proposed architecture may be realized with varying level of parallelism, depending on the parameter p. As follows from Eq. (11) the time period complexity of the implementation varies between $T_L(C1) = 2^{m-J}$ and $T_1(C1) = L2^{m-J}$. Thus, the throughput of the architecture is $2^J/L$ to 2^J times faster than that of the fastest known architectures. The possibility of realization of the architecture with a varying level of parallelism also gives an opportunity to trade–off the time and hardware complexities. It also should be noted that the architecture is very regular and needs an easy control (which is, essentially, a clock only) unlike e.g., the architecture of [17]. It does not contain a feedback, a switch, or long connections dependent on the size of the input but only connections of the maximum of $O(L)$ length. Thus, it can be implemented as a semisystolic array.

B. Type 2 Core DWT Architecture

When implementing the basic operations (6) and (7) of Algorithm 3.1, multiplicands needed for the time unit $n = 1, \ldots, L_p - 1$ within the branch $i = 0, \ldots, 2^{J-j} - p/2 - 1$ can be obtained from the results obtained at step $n - 1$

---

* The efficiency or hardware utilization is $E = (T(1) \cdot 100\%) / (K \cdot T(K))$, where $T(1)$ and $T(K)$ are the time complexities with one PE and with K PEs, respectively.

within the branch $i + p/2$. With this, the following modification of the basic algorithm may be derived. Denote

$$l'_k = \begin{cases} l_k \text{ for } k = 0, \dots, p-1 \\ l_k / l_{k-p,} \text{ for } k = p, \dots, L-1 \end{cases};$$

$$h'_k = \begin{cases} h_k \text{ for } k = 0, \dots, p-1 \\ h_k / l_{k-p,} \text{ for } k = p, \dots, L-1 \end{cases}$$

Algorithm 3.2

1. For $s = 0, \dots, 2^{m-J} - 1$ set $x_{LP}^{(0,s)} = x(s \cdot 2^J : (s+1) \cdot 2^J - 1)$;
2. For $s = s*(1), \dots, 2^{m-J} + s*(J) - 1$, For $j = J_1, \dots, J_2$ do in parallel
2.1. Set $\hat{x}^{(j-1,s-s*(j))}$ according to (4)
2.2. For $i = 0, \dots, 2^{J-j} - 1$ do in parallel
For $k = 0, \dots, p-1$
{set $z(i, 0, k) = l_k \hat{x}^{(j-1,s-s*(j))}(2i + k)$;
Compute $S_{LP}(i) = \sum_{k=0}^{p-1} z_{LP}(i, 0, k)$; $\quad S_{HP}(i) = \sum_{k=0}^{p-1} z_{HP}(i, 0, k)$; }
For $n = 1, \dots, L_p - 1$ do sequentially
For $k = 0, \dots, p-1$

$$\text{set } z_{LP}(i, n, k) = \begin{Bmatrix} l'_{np+k} z(i + p/2, n-1, k) \text{ if } i < 2^{J-j} - p/2 \\ ; \\ l_{np+k} \hat{x}^{(j-1,s-s*(j))}(2i+k) \text{ if } i \geq 2^{J-j} - p/2 \end{Bmatrix};$$

$$\text{set } z_{HP}(i, n, k) = \begin{Bmatrix} h'_{np+k} z(i + p/2, n-1, k) \text{ if } i < 2^{J-j} - p/2 \\ h_{np+k} \hat{x}^{(j-1,s-s*(j))(2i+k)} \text{ if } i \geq 2^{J-j} - p/2 \end{Bmatrix}$$

$$\text{Compute } S_{LP}(i) = S_{LP}(i) + \sum_{k=0}^{p-1} z_{LP}(i, n, k);\ S_{HP}(i) = S_{HP}(i) + \sum_{k=0}^{p-1} z_{HP}(i, n, k);$$

Set $x_{LP}^{(j,s-s*(j))}(i) = S_{LP}(i)$; $x_{HP}^{(j,s-s*(j))}(i) = S_{HP}(i)$
3. Form the output vector (see Step 3 of Algorithm 2).

The general structure of Type 2 core DWT architecture is presented in Fig. 4, where now the dashed lines that show connections between PEs of one stage are valid. Except for p inputs and two outputs (later on called main inputs and main outputs), every PE now has additional p inputs and p outputs (later on called intermediate inputs and outputs). The p intermediate outputs of $PE_{j,i+p/2}$ are connected to the p intermediate inputs of $PE_{j,i}$, $i = 0, \dots, 2^{J-j} -$

$p/2 - 1$. Other connections within Type 2 core DWT architecture are similar to those within Type 1 core DWT architecture.

Functionalities of the blocks of Type 2 core DWT architecture are also similar to those of Type 1 core DWT architecture. The difference is only in the functionality of PEs which at every time unit $n = 0, \ldots, L_p - 1$ of every operation step is to compute two inner products of a vector, say, x, on its p either main or intermediate inputs with two vectors of predetermined coefficients, say LP' and HP' of length p as well as to compute a point-by-point product of x with LP'. At the time unit $n = 0$ the vector x is the one formed on the main p inputs of the PE and at time units $n = 1, \ldots, L_p - 1$ it is the one formed on the intermediate inputs of the PE. Results of both inner products computed during one operation step are accumulated and passed to the two main outputs of the PE while the results of the point-by-point products are passed to the intermediate outputs of the PE. A possible structure of PEs for Type 2 core DWT architecture for the case of $p = 2$ is presented in Fig. 6. Structures for arbitrary p and for $p = 1$, $p = 2$, and $p = L_{max}$, can be easily designed, similar to those in Fig. 5.

Similar to the case of Type 1 core DWT architecture, one can see that Type 2 core DWT architecture implements Algorithm 3.2 with time delay and time period characteristics given by Eqs. (10) and (11). The other characteristics of these two architectures are also similar. In particular, it is very fast and it may be implemented as a semisystolic architecture and with varying level of parallelism giving opportunity of trade-off between time and hardware complexities. The difference between these two architectures is that the shift registers of data routing blocks of Type 1 core DWT architecture are replaced with additional connections between PEs within Type 2 core DWT architecture.

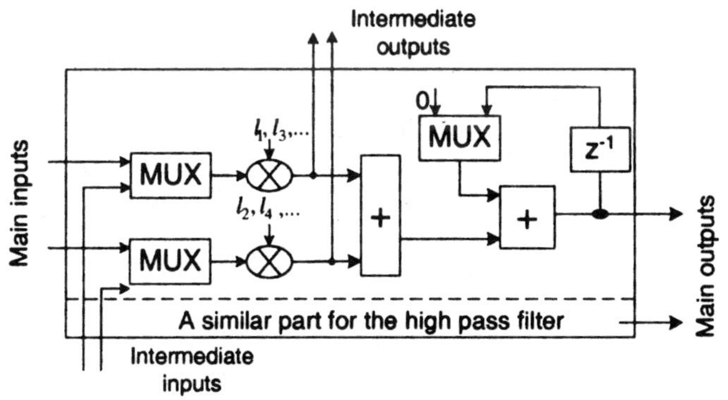

Figure 6 A possible realization of a PE for the Type 2 core DWT architecture; $p = 2$.

C. Multicore DWT Architectures

The two types of core DWT architectures described above may be implemented with a varying level of parallelism depending on the parameter p. Further flexibility in the level of parallelism is achieved within multicore DWT architectures by introducing a new parameter $r = 1, ..., 2^{m-J}$. The multicore DWT architecture is, in fact, obtained from corresponding (single) core DWT architecture by expanding it r times. Thus, one can again consider Fig. 4 for the general structure of multicore architectures but the numbers of PEs at every pipeline stage should be multiplied by r.

The two types of multicore DWT architectures are r times faster than the (single)core DWT architectures, that is, a linear speed-up with respect to parameter r is achieved:

$$T_d(C1) = \left(2^{m-J} + \hat{s}(J)\right)\lceil L/p \rceil / r \tag{12}$$

time units and the throughput or the time period is equal to

$$T_p(C1) = 2^{m-J}\lceil L/p \rceil / r \tag{13}$$

time units. Thus further speed-up and flexibility for trade-off between time and hardware complexities is achieved within multicore DWT architectures. Architectures are modular and regular and may be implemented as semi-systolic arrays. As an example of the multicore DWT architecture for the case of $p = L = L_{\max}$ and $r = 2^{m-J}$ one can consider the DWT flowgraph itself (see Fig. 2), where nodes (rectangles) should be considered as PEs and small circles as latches. This example of realization was reported in [29–30], where it was referred to as fully parallel pipelined (FPP) architecture.

D. Variable Resolution DWT Architectures

The above-described architectures implement DWTs with the number of octaves not exceeding a given number J. They may implement DWTs with smaller than J number of octaves though with some loss in hardware utilization. The variable resolution DWT architecture implements DWTs with arbitrary number J' of octaves whereas the efficiency of the architecture remains approximately 100% whenever J' is larger than or equal to a given number $J_{\min}$.

The general structure of the variable resolution DWT architecture is shown in Fig. 7a. It consists of a core DWT architecture corresponding to $J_{\min}$ DWT octaves and an arbitrary serial DWT architecture, for, instance, an RPA-based one [14–17], [19 20], [22]. The core DWT architecture implements the first $J_{\min}$ octaves of the J'-octave DWT. The low-pass results from the *out(0)* of the core DWT architecture are passed to the serial DWT architec-

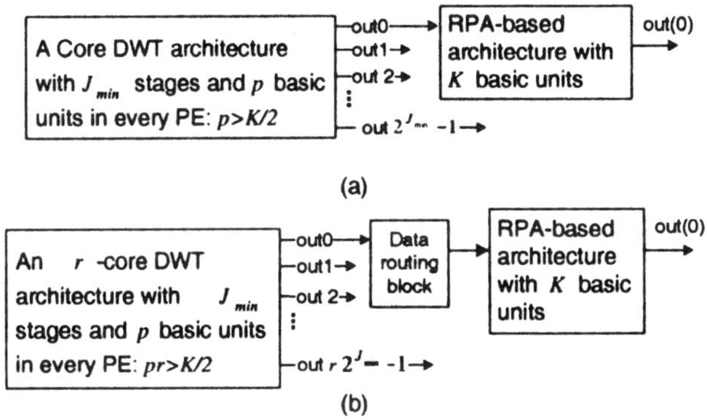

Figure 7 The variable resolution DWT architecture: (a) based on a single core DWT architecture; (b) based on a multicore DWT architecture.

ture. Then the serial DWT architecture implements the last $J' - J_{\min}$ octaves of the J'-octave DWT. Since the core DWT architecture may be implemented with a varying level of parallelism it can be balanced with the serial DWT architecture in such a way that approximately 100% of hardware utilization is achieved whenever $J' \geq J_{\min}$.

To achieve the balancing between the two parts, the core DWT architecture must implement a $J_{\min}$-octave N-point DWT with the same throughput or faster as the serial architecture implements $(J' - J_{\min})$-octave M-point DWT ($M = (N/2^{J_{\min}})$). Serial architectures found in the literature implement an M-point DWT, either in $2M$ time units [14,15], or in M time units [14–19], correspondingly employing either L or $2L$ basic units (BUs, multiplier-adder pairs). They can be scaled down to contain an arbitrary number $K \leq 2L$ BUs so that an M-point DWT would be implemented in $M\,[2L/K]$ time units. Since the (Type 1 or Type 2) core DWT architecture implements a $J_{\min}$-octave N-point DWT in $N\,[L/p]\,/2^{J_{\min}}$ time units the balancing condition becomes $[L/p] \leq [2L/K]$, which will be satisfied if $p = [K/2]$. With this condition the variable resolution DWT architecture will consist of totally A BUs,

$$A = 2p\left(2^{J_{\min}} - 1\right) + K = \begin{cases} K2^{J_{\min}}, \text{ if } K \text{ is even} \\ (k+1)2^{J_{\min}} - 1, \text{ if } K \text{ is odd} \end{cases}$$

and will implement a J'-octave N-point DWT in T_d time units, where

$$T_d = N\,\lceil 2L/K \rceil\,/2^{J_{\min}}$$

A variable resolution DWT architecture based on a multicore DWT architecture may also be constructed (see Fig. 7b), where now a data routing block is inserted between the multicore and serial DWT architectures. The functionality of the data routing block is to accept in parallel and to output serially digits at the rate of r samples per operation step. The balancing condition in this case is $rp = \lceil K/2 \rceil$. The area–time characteristics are similar to those for the single core based architectures.

IV. CONCLUSIONS AND A SUMMARY OF THE PERFORMANCE

Table 1 presents a comparative performance of the proposed architectures with some conventional architectures. In this table, as it is commonly accepted in the literature, the area of the architectures was counted as the number of used multiplier-adder pairs which are the basic units in DWT

Table 1 Comparative Performance of Some DWT Architectures

Architecture	Area. A (No. of BUs)	Period, T_p	AT_{p}^2
Architectures in [14], [15]	L	$2N$	$4N^2L$
Architectures in [14]–[19]	$2L$	N	$2N^2L$
lifting-based [11] (specific filters)	≈ 4	$\alpha(L)N\sum_{j=1}^{J}2^{1-j}$, $\alpha(L)=1,2$	$\approx 4N^2$ or $\approx 16N^2$
Architectures in [12],[24]	JL	N	$\mathrm{J}N^2L$
Architectures of [27],[28]	$4L$ or $\sum_{j=1}^{J}\lceil L/2^{j-2}\rceil$	$N/2$	$\approx N^2L$
FPP DWT [29]–[30] (pipelined)	$2NL(1-1/2^J)$	1 (per vector)	$2NL(1-1/2^J)$
LPP DWT [29]–[30]	$2L\ (2^J-1)$	$N/2^J$	$N^2L(2^J-1)/2^{2J-1} \approx N^2L/2^{J-1}$
Single core DWT (Type 1 or 2)	$2p\ (2^J-1)$	$N\lceil L/p\rceil/2^J$	$\approx N^2p(L/p)^2\ /2^{J-1}$
Single core DWT, p = 1	$2\ (2^J-1)$	$NL/2^J$	$\approx N^2L^2/2^J$
Single core DWT p = L_{max} ($L \le L_{\max}$)	$2L_{\max}\ (2^J-1)$	$N/2^J$	$\approx N^2L_{\max}/2^{J-1}$
Multicore DWT	$2pr\ (2^J-1)$	$(N\lceil L/p\rceil)\ /\ (r2^J)$	$\approx (N^2p\lceil L/p\rceil^2)\ /\ (r2^{J-1})$
Multicore DWT, r = 4, p = 1	$8\ (2^J-1)$	$NL/2^{J+2}$	$\approx N^2L^2/2^{J+1}$
Multicore DWT r = 4, p = L_{max} ($L \le L_{\max}$)	$2rL_{\max}\ (2^J-1)$	$N/(r2^J)$	$\approx (N^2L_{\max})\ /\ (r2^{J-1})$
Variable resolution single core DWT $p \ge \lceil K/2 \rceil$, ($K \le 2L$)	$2p\ (2^{J\min}-1) + K \approx K2^{J\min}$	$N\lceil 2L/K\rceil\ /2^{J\min} \approx 2NL(K2^{J\min})$	$\approx \frac{N^2L^2}{K2^{J\min-2}}$

architectures. The time unit is counted as the time period of one multiplication since this is the critical pipeline stage. Characteristics of the DWT architectures proposed in this chapter (the last seven rows in Table 1), are given for arbitrary realization parameters $L_{\max}$, p, and r as well as for some examples of parameter choices. It should be mentioned that the numbers of BUs used in the proposed architectures assume the PE examples of Figs 5 and 6 (where PE with p inputs contains $2p$ BUs). However, PEs could be further optimized to involve a lower number of BUs.

As follows from Table 1, the proposed architectures, compared to the conventional ones, demonstrate excellent time characteristics at moderate area requirements. Advantages of the proposed architectures are best shown when considering the performances with respect to AT_p^2 criterion, which is commonly used to estimate performances of high-speed oriented architectures. Architectures presented in the first two rows of Table 1 are either nonpipelined or restricted (only two-stage) pipelined ones and they operate at approximately 100% hardware utilization as is the case for our proposed architectures. So their performance is "proportional" to the performance of our architectures, which, are however, much more flexible in the level of parallelism resulting in a wide range of time and area complexities.

The best performance was achieved with the lifting-based architecture [11] (the third row of Table 1), which is specifically optimized for filters used in the JPEG2000 image compression standard and gives slightly better than "proportional" performance as compared to the proposed architectures. It should be noted, however that this architecture is also nonpipelined and utilizes PEs extensively optimized for the specific set of filters (in particular, the lifting scheme utilizes the correlation between the low- and high-pass filter coefficients). Actually, similar PEs may also be used in the proposed architectures, resulting in less area requirements. The fourth row of Table 1 presents J-stage pipelined architectures with a poor hardware utilization and consequently a poor performance. The fifth and sixth rows of Table 1 present architectures from our previous publications, which are J-stage pipelined and achieve 100% hardware utilization and good performance but do not allow a flexible range of area and time complexities as the architectures proposed in this chapter.

REFERENCES

1. Mallat, S. G. (1989). A theory for multiresolution signal decomposition; the wavelet representation. *IEEE Trans Pattern Analysis and Machine Intelligence* 2:674–693.
2. Vetterli, M., Kovacevic, J. (1995). *Wavelets and Subband Coding*. New York: Prentice-Hall.

3. Daubachies, I. (1992). *Ten Lectures on Wavelets*. Philadelphia (PA): SIAM.
4. Sweldens, W. (1995). The lifting scheme: a New philosophy in biorthogonal wavelets constructions. Proceedings of SPIE International Conference on Wavelet Applications in Signal and Image Processing III, San Diego, Vol. 2569. pp 68–79.
5. Beylkin, G., Coifman, R., Rokhlin, V. (1992). Wavelets in numerical analysis. *Wavelets and their Applications*. New York: Jones and Bartlett, pp. 181–210.
6. Jiang, W., Ortega, A. (2001). Lifting factorization-based discrete wavelet transform architecture design. *IEEE Trans Circ Syst Video Tech* 11:651–657.
7. Diou, C., Torres, L., Robert, M. (2000). A wavelet core for video processing. Presented at the IEEE International Conference on Image Processing, Vancouver.
8. Lafruit, G., Nachttegraele, L., Bormans, J., Engles, M., Bolsens, I. (1999). Optimal memory organization for scalable texture codecs in MPEG-4. *IEEE Trans on Circ Syst Video Tech* 9:218–243.
9. Ferretti, M., Rizzo, D. (2001). A Parallel architecture for the 2-D discrete wavelet transform with integer lifting scheme. *J VLSI Signal Proc* 28:165–185.
10. Huang, Ch-T., Tseng, Po-Ch., Chen, L-G. (2002). Efficient VLSI architectures of lifting-based discrete wavelet transform by systematic design method. Proceedings of the IEEE International Symposium on Circuits and Systems, Phoenix, pp. 565–568.
11. Andra, K., Chakrabarti, Ch., Acharya, T. (2002). A VLSI architecture for lifting-based forward and inverse wavelet transform. *IEEE Trans Sign Proc* 50: 966–977.
12. Pan, S.B., Park, R.H. (1997). New systolic arrays for computation of the 1_D discrete wavelet transform. Proceedings of the IEEE International Conference on Acoustics, Speech, and Signal Processing, Munich, pp. 113–4116.
13. Premkumar, A.B., Madhukumar, A.S. (1997). An efficient VLSI architecture for the computation of 1-D discrete wavelet transform. Proceedings of the IEEE International Conference on Information, Communications and Signal Processing, Singapore, pp. 1180–1184.
14. Vishwanath, M., Owens, R. M., Irwin, M. J. (1995). VLSI architectures for the discrete wavelet transform. *IEEE Trans Circ Syst-II: Analog Dig Sign Proc* 42:305–316.
15. Fridman, J., Manolakos, S. (1997). Discrete wavelet transform: data dependence analysis and synthesis of distributed memory and control array architectures. *IEEE Trans Sign Proc* 45:1291–1308.
16. Parhi, K., Nishitani, T. (1993). VLSI architectures for discrete wavelet transforms. *IEEE Trans VLSI Syst* 1:191–202.
17. Denk, T.C., Parhi, K. (1998). Systolic VLSI architectures for 1-d discrete wavelet transform. Proceedings of the Thirty-Second Asilomar Conference on Signals, Systems and Computers, Monterey, pp. 1220–1224.
18. Knowles, G. (1990). VLSI architecture for the discrete wavelet transform. *El Let* 26:1184–1185.
19. Chakrabarti, Ch., Vishwanath, M. (1995). Efficient realizations of discrete and

continuous wavelet transforms: from single chip implementations to mappings on SIMD array computers. *IEEE Trans Sign Proc* 43:759–771.
20. Chakrabarti, Ch., Vishwanath, M., Owens, R. M. (1996). Architectures for wavelet transforms: a survey. *J VLSI Sign Proc* 14:171–192.
21. Grzeszczak, A., Mandal, M. K., Panchanatan, S. (1996). VLSI implementation of discrete wavelet transform. *IEEE Trans VLSI Systems* 4:421–433.
22. Vishwanath, M., Owens, R.M. (1996). A common architecture for the DWT and IDWT. Proceedings of the IEEE International Conference on Application Specific Systems, Architectures and Processors, San Diego, pp. 193–198.
23. Yu, C., Hsieh, C.H., Chen, S.J. (1997). Design and implementation of a highly efficient VLSI architecture for discrete wavelet transforms. Proceedings of the IEEE International Conference on Custom Integrated Circuits, Santa Clara (CA), pp 237–240.
24. Syed, S. B., Bayoumi, M. A. (1995). A scalable architecture for discrete wavelet transform. Proceedings of Computer Architectures for Machine Perception, Italy: Como, pp. 44–50.
25. Marino, F. (2000). A 'Double-Face' bit-serial architecture for the 1-D discrete wavelet transform. *IEEE Trans Circ Syst-II: Analog Dig Sign Proc* 47:65–71.
26. Vishwanath, M. (1994). The recursive pyramid algorithm for the discrete wavelet transform. *IEEE Trans Sign Proc* 42:673–677.
27. Marino, F., Gevorkian, D., Astola, J. (2000). Highly efficient high-speed/low-power architectures for the 1-D DWT. *IEEE Trans Circ Syst-II: Analog Dig Sign Proc* 47:1492–1502.
28. Marino, F., Gevorkian, D., Astola, J. (2000). High-speed/low-power 1-D DWT architecture with high efficiency. Proceedings of the IEEE International Conference on Circuits and Systems, Geneva, pp. 337–340.
29. Gevorkian, D., Marino, F., Agaian, S., Astola, J. (2000). Highly efficient fast architectures for discrete wavelet transforms based on their flowgraph representation. Proceedings of the European Signal Processing Conference, Tampere (Finland).
30. Gevorkian, D., Marino, F., Agaian, S., Astola, J. (2000). Flowgraph representation of discrete wavelet transforms and wavelet packets for their efficient parallel implementation. Proceedings of TICSP International Workshop on Spectral Transforms and Logic Design for Future Digital Systems, Tampere (Finland).

4

Stride Permutation Access in Interleaved Memory Systems

Jarmo Takala and Tuomas Järvinen
Tampere University of Technology, Tampere, Finland

Over the years several techniques have been proposed to increase data transfer rates between memory and computational resources in processor architectures. This performance bottleneck can be avoided by allowing several simultaneous accesses to the memory, which implies that the memory system should have several ports. Multiported memories can be used but they are an expensive solution especially when the number of ports is large.

A more area-efficient method is to use several independent memory banks or modules, which can be accessed in parallel. The principal problem in such memory systems is to distribute data over multiple modules in such a way that parallel access is possible. However, there is no general purpose solution to the distribution problem and several methods have been proposed, which assume that parallel accesses are most likely to be made subsections of data arrays.

This chapter focuses on parallel access with a specific access order, stride permutation, which has several applications in the field of digital signal processing (DSP). The organization of the remainder of this chapter is, in Section I, we describe two principal types of interleaved memory systems: time and space-multiplexed systems. Access schemes are discussed in Section II. The principles of low-order interleaving, row rotation, and linear transformation are provided. Section III considers one specific access pattern, stride access, which is one of the basic access patterns discussed in several research papers. Requirements for supporting multiple strides in a general case is given. In Section IV, stride permutation access is described with the aid of the definition of stride permutation. Some of its properties are given and motivation for developing an access scheme for this access pattern is provided by describing some of its applications. In Section V, a conflict-free access scheme for stride permutation access is presented. The used assumptions are given, the scheme is described in detail and validated through simulations, and an implementation is presented. A summary is provided in Section VI.

I. INTERLEAVED MEMORY SYSTEMS

One well-established technique for increasing the data transfer rate between memory and computational resources is memory interleaving where data is distributed over multiple independent memory modules. In general, such memory systems exploit either time or space multiplexing [1]. *Time-multiplexed memories* are used in vector machines to match the processor cycle time and memory access time. Memory accesses will require t cycles to complete, and this delay is hidden by sending t access requests to the memory system over a single bus at consecutive cycles. Each request is sent into a different memory module. If the operands lie in the same memory modules, the next access can be performed after t cycles. The principal block diagram of a time-multiplexed memory system can be seen in Fig. 1a. *Space-multiplexed memories* are used in SIMD processing (i.e., several access requests are sent to the memory system over multiple buses, thus the memory latency is not hidden). The memory system requires an interconnection network to provide a communication path from processing units to different memory modules. The principal block diagram of a space-multiplexed memory system is illustrated in Fig. 1b.

In both the previous systems, the memory bandwidth is increased by allowing several simultaneous memory accesses to be directed to different memory modules. If Q accesses can be distributed over Q modules such that all the modules are referenced, Q-fold speedup can be achieved. Unfortu-

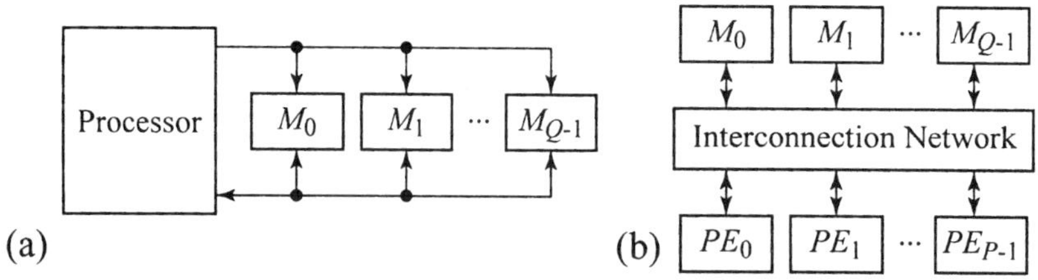

Figure 1 Interleaved memory systems: (a) time-multiplexed and (b) space-multiplexed. (M_k: memory module. PE_k: processing element.)

nately the operands to be accessed in parallel often lie in the same memory module thus the parallel access can not be performed resulting in performance degradation. Such a situation is referred to as a conflict. The principal problem in interleaved memory systems is to find a method to distribute data over the memory modules in such a way that conflicts are avoided. For this research problem, a traditional assumption has been that the parallel accesses are most likely to be made to subsections of matrices (e.g., rows, columns, or diagonals). Several methods suggest that the number of memory modules Q is larger than the processing elements P (i.e., number of simultaneous accesses). In this chapter, we consider only *matched memory systems* where the number of memory modules equals the number of processing elements, $P = Q$.

While the memory interleaving has gained popularity in the supercomputer area, it has received a little consideration in embedded systems. In general, studies in supercomputing are based on static stream models (i.e., assumption that the memory accesses are most likely to access a section of a data array). However, in real-time embedded systems, the data streams are affected by dynamic behaviors, (e.g., program execution, cache behavior, and access scheduling). Such a characteristic is considered in approach, reported in [2], where the memory organization is optimized before scheduling and synthesis of data paths and controllers. The method minimizes the number of memory modules, may use different word widths in each module, and results in a memory organization containing parallel memory modules but those are not necessarily interleaved. A design method for embedded systems based on interleaved memories is reported in [3]. In principle, the objective is to improve data access locality by applying several transformations to the given application (e.g., for data layout and loops). Next the number of memory modules is estimated and the application can be compiled to the resulting architecture. In [4], a memory synthesis method for interleaved memories is proposed. The method is targeted to application-specific processors, thus the order of memory accesses is known before and the address generators can be optimized for the given accesses.

II. ACCESS SCHEME

The method of distributing data over modules is referred to as an *access scheme*, which is a function mapping addresses into storage locations. When an N-element array is distributed over Q memory modules, an access scheme performs two mappings; it maps a $\lceil \log_2 N \rceil$-bit address $a = (a_{n-1}, a_{n-2}, \ldots, a_0)^T$ into a $\lceil \log_2 Q \rceil$-bit *module address*, m, and into a *row address*, r, defining the storage location in the selected memory module.

The most simple access scheme is to obtain row and module addresses by extracting fields from the address a, i.e.,

$$r = \lfloor a/Q \rfloor \tag{1}$$

$$m = a \bmod Q \tag{2}$$

Such a scheme, *low order interleaving*, is illustrated in Fig. 2a. This scheme performs well in linear access but the performance is degraded when other types of access patterns are used [5].

In order to support a larger set of access patterns, the *row rotation* (alternatively *skewed*) scheme was introduced in [6]. Formally, address mapping can be described as follows

$$r = \lfloor a/Q \rfloor \tag{3}$$

$$m = (a + \lfloor a/N \rfloor) \bmod Q \tag{4}$$

When $Q = 2^q$, the module address is formed simply by extracting two $\log_2 Q$-bit fields from the address a and adding the fields together as shown in Fig. 2b.

Often a prime number of memory modules is used since it typically results in a larger set of conflict-free access patterns. The inflexibility of the traditional row rotation schemes is illustrated by the following theorem [7].

Theorem 1. *An $N \times N$ matrix, $N = 2^n$, cannot be stored into N memory modules by any row rotation scheme such that all the rows, columns, and diagonals can be accessed conflict-free.*

The prime number of modules implies that the address computation needs a modulo operation of a number, which is not a power-of-two. Such an operation requires large circuitry. Furthermore, prime number of memory modules often results in low memory utilization, (i.e., not all the memory locations are allocated) [8].

In [9], row rotation scheme was generalized as a periodic storage scheme, which supports irregular and overlapped access patterns. A row rotation scheme supporting power-of-two number of memory modules was proposed in [10], where the principal idea was to partition the scheme into several sub schemes (i.e., a different sub scheme is applied to each part of the entire data vector). This results in a need to support several schemes instead of a single scheme.

In [11], a scheme was introduced where the address mapping is a *linear transformation* based on modulo-2 arithmetic. This implies that the arithmetic is realized with bit-wise exclusive-OR (XOR) operations, thus modulo operations are not needed and carry delay of adders used in row rotation scheme is avoided. Linear transformation schemes are often called as *XOR*

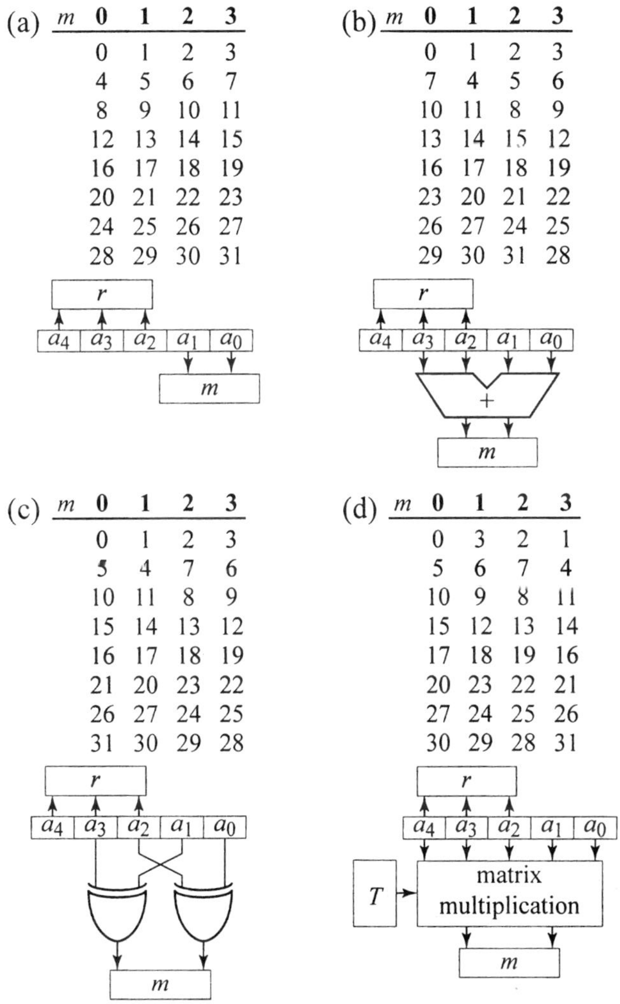

Figure 2 Examples of access schemes for a 32-element vector on a 4-module system: (a) low order interleaving; (b) row rotation; and linear transformations according to method (c) in [5] and (d) in [14].

schemes. The address mappings in linear transformation can be expressed with binary transformation matrices as

$$r = Ka \tag{5}$$

$$m = Ta \tag{6}$$

It should be noted that in this representation the least significant bit of a is in the bottom of the vector. Matrices K and T are the row and module transformation matrix, respectively.

Often K consists of ones in the main diagonal thus the row address r is obtained simply by extracting the $(n - q)$ most significant bits of the address a, i.e.,

$$r = (a_{n-1}, a_{n-2}, \ldots, a_q)^T \tag{7}$$

The module transformation matrix T is often expressed in the following form

$$m = Ta = (T_H | T_L)a \tag{8}$$

where T_L is the rightmost $q \times q$ square matrix in T and T_H is the remaining $q \times (n - q)$ matrix in T. An example of linear transformation is depicted in Fig. 2c and the corresponding matrix T is

$$T = \begin{pmatrix} 0 & 1 & 0 & 1 & 0 \\ 0 & 0 & 1 & 0 & 1 \end{pmatrix} \tag{9}$$

In general, linear transformations have two advantages over row rotation schemes: the computation of module address is independent of the number of memory modules and the scheme has flexibility in performing address mappings [5]. These schemes have been analyzed in several papers and the basic requirement for the data distribution was derived in [12] as follows.

Theorem 2. An interleaved memory system has a unique storage location for each addressed element iff the matrix T_L has full rank.

In [13], it was suggested that T should have full rank and, in particular, the main diagonal of T should consist of 1s. Missing 1s in the main diagonal may result in poor performance for linear access. In addition, off-diagonal 1s complicate the construction of address generators.

III. STRIDE ACCESS

One specific, often used access pattern is *stride access* where indices in consecutive accesses differ by a constant S resulting in a sequence of addresses $(i, i + S, i + 2S, i + 3S, \ldots, i + (S - 1)S)$ for some starting address i. When such an access is performed in parallel, every Sth element of an array is accessed concurrently. Stride accesses occur often in application programs, especially in matrix computations (e.g., when accessing rows and columns of a matrix). It is also often used in image processing where the image data is accessed in the form of rectangles, grids, or chessboards. When a vector $x = (x_0, x_1, \ldots)^T$ is accessed with stride S in a system containing Q memories and processing elements, a single parallel access is referencing to the elements $(x_i, x_{i+S}, x_{i+2S}, \ldots, x_{i+(Q-1)S})^T$.

In [14], a linear transformation for matched systems is reported, which supports several power-of-two strides. The proposed module transformation matrix forms a recursive pattern of repeating triangles. Such a matrix can be generated with a recursive rule: each element is XOR of its neighbors to the right and above. For example, when mapping a 32-element array over four modules, the module transformation matrix T is the following:

$$T = \begin{pmatrix} 1 & 1 & 1 & 1 & 1 \\ 1 & 0 & 1 & 0 & 1 \end{pmatrix} \tag{10}$$

The contents of the memories in this case are illustrated in Fig. 2d. The proposed realization of the module address generator is based on matrix multiplication as seen in the Fig. 2d. The implementation is complex, especially if several array lengths need to be supported.

A linear transformation scheme supporting linear and single stride accesses is reported in [13]. Conflict-free stride access can be performed for any array length and any initial address. The implementation is extremely simple requiring only bit-wise XOR operations and a shifter for address field extraction. Support for several strides is considered in [5] and strides of form $a2^s$ are supported. However, the number of memory modules needs to be greater than the number of parallel accesses. In [15], linear transformations in matched and unmatched systems were discussed.

In [5], stride accesses were investigated with the aid of transformation periodicity referring to the minimum period of the sequence of module numbers generated when consecutive addresses were used as the input sequence. This results in the following requirement.

Theorem 3. In matched memory system, $S = 2^s$ stride access over $Q = 2^q$ memories is conflict-free iff the linear transformation matrix T is

a) periodic SQ and

b) $(a + iS)T \bmod Q = (a + jS)T \bmod Q$ iff $i \bmod Q = j \bmod Q$.

Condition (a) guarantees that the access is conflict-free regardless of the array length and initial address of the array. Condition (b) defines that each memory module is referenced only once in Q parallel accesses. It can be shown that under the previous constraints a conflict-free access scheme supporting several strides cannot be designed [13].

IV. STRIDE PERMUTATION ACCESS

Stride permutations can be described with the aid of matrix transpose; stride-by-S permutation of an N-element vector can be performed by dividing the vector into S-element subvectors, organizing them into $S \times (N/S)$ matrix form, transposing the obtained matrix, and rearranging the result back to the vector representation [16]. This interpretation implies that the stride S must be a factor of vector length, i.e., N rem $S = 0$ where rem denotes remainder after division. Another interpretation is to use indexing functions as used in the following formal definition.

Definition 1 (Stride permutation). Let us assume a vector $X = (x_0, x_1, \ldots, x_{N-1})^T$. *Stride-by-S permutation reorders X as vector Y,* $Y = (x_{f_{N,S}(0)}, x_{f_{N,S}(1)}, \ldots, x_{f_{N,S}(N-1)})^T$, *where the index function* $f_{N,S}(i)$ *is given as*

$$f_{N,S}(i) = (iS \bmod N) + \lfloor iS/N \rfloor;$$
$$N \text{ rem } S = 0 \quad i = 0, 1, \ldots, N-1 \tag{11}$$

where $\lfloor \cdot \rfloor$ *is the floor function.*

The stride permutation can also be expressed in matrix form as $Y = P_{N,S}X$ where $P_{N,S}$ is stride-by-S permutation matrix of order N defined as

$$[P_{N,S}]_{mn} = \begin{cases} 1, \text{ iff } & n = (mS \bmod N) + \lfloor mS/N \rfloor \\ 0, & \text{otherwise} \end{cases}$$
$$m,\ n = 0, 1, \ldots, N-1 \tag{12}$$

For example, the permutation matrix $P_{8,2}$ associated to stride-by-2 permutation of an 8-element vector is the following (blank entries represent zeroes):

$$P_{8,2} = \begin{pmatrix} 1 & & & & & & & \\ & & 1 & & & & & \\ & & & & 1 & & & \\ & & & & & & 1 & \\ & 1 & & & & & & \\ & & & 1 & & & & \\ & & & & & 1 & & \\ & & & & & & & 1 \end{pmatrix}$$

In this chapter, we limit ourselves to practical cases where the stride and array lengths are powers-of-two, $N = 2^n$, $S = 2^s$. Some properties of stride permutations in such cases are in the following.

Theorem 4. (Factorization of stride permutations). Let $ab \leq N$, then

$$P_{N,ab} = P_{N,a}P_{N,b} = P_{N,b}P_{N,a} \tag{13}$$

The proof for the previous theorem can be found (e.g., from [17]).

Corollary 1 (Periodicity). *Stride permutations are periodic with the following properties.*

(1) Period of $P_{2^n,2^s}$ is $\mathrm{lcm}(n,s)/s$ where $\mathrm{lcm}(a,b)$ denotes the least common multiple of n and s. In other words,

$$I_{2^n} = \prod_{1}^{\mathrm{lcm}(n,s)/s} P_{2^n,2^s} \tag{14}$$

(2) Consecutive stride permutations always result in a stride permutation:

$$P_{2^n,2^a}P_{2^n,2^b} = P_{2^n,2^{(a+b) \bmod n}} \tag{15}$$

Proof. Property (2) If $a + b > n$, the left side of Eq. (15) can be written as $P_{2^n,2^{kn+(a+b) \bmod n}} = P_{2^n,2^{kn}}P_{2^n2^{(a+b) \bmod n}}$ where $k > 1$ is an integer. By substituting 2^n for S in Eq. (11), we find that $P_{2^n,2^n} = P_{2^n,1} = I_{2^n}$. Therefore, $P_{2^n,2^{kn}}$ $P_{2^n,2^{kn}} = I_{2^n}$ and the result follows. Property (1) Let us assume that period of $P_{2^n,2^s}$ is k, thus $ks \bmod n = 0$, i.e., ks is a multiple of n. This implies that k is a multiple of n/s, i.e., $k = mn/s$. k has to be integer, thus s has to be a factor of mn. The smallest number fulfilling the requirement is $\mathrm{lcm}(n,s)$ and, therefore, $k = \mathrm{lcm}(n,s)/s$.

Stride permutations have several practical applications. For example, the previously discussed matrix transpose interpretation of stride permutation implies that an $N \times N$ matrix in a vector form can be transposed by reordering the N^2-element vector according to stride-by-N permutation, $P_{N^2,N}$. Therefore, an $N \times N$ matrix stored into a memory array can be transposed by accessing its elements in stride-by-N order.

The well-known perfect shuffle permutation is a special case of stride permutation: stride-by-$N/2$ permutation of an N-point sequence, $P_{N,N/2}$. Perfect shuffle has close relation to several practical algorithms; e.g., Cooley-Tukey radix-2 fast Fourier transform (FFT) algorithm can be scheduled into a form where the interconnections between the processing columns of the signal flow graph are perfect shuffles. Radix-2 algorithms can also be derived into a form where the topology is according to stride-by-2 permutation as illustrated in Fig. 3a. In radix-4 algorithms, the interconnections can be stride-by-4 permutations as depicted in Fig. 3b. Fast algorithms for other discrete trigonometric transforms with corresponding topology exist, (e.g., for discrete sine, cosine, and Hartley transforms) [18,19].

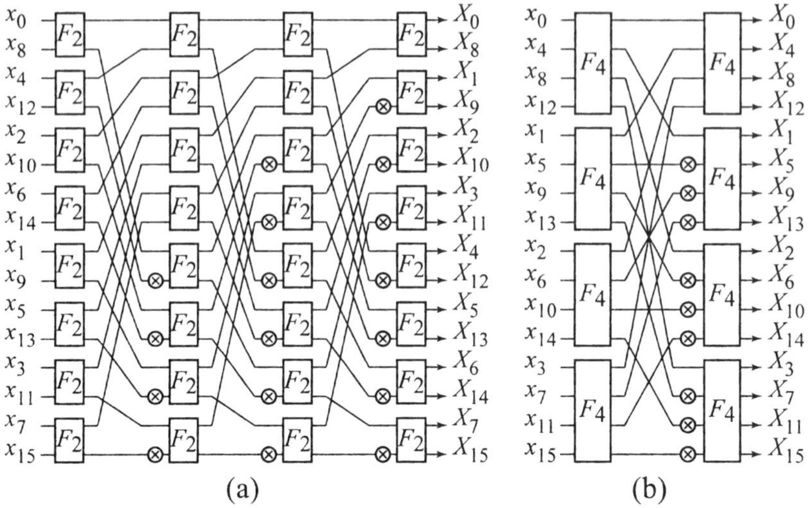

Figure 3 Signal flow graphs of FFT algorithms: (a) radix-2 and (b) radix-4 algorithm. (F_k: k-point FFT.)

Stride permutations can also be found in trellis coding and especially in Viterbi algorithms used for decoding convolutional codes. Convolutional encoders are often described with the aid of a shift register model (illustrated in Fig. 4). The state of the encoder X_t at a given time instant t is defined by the contents of the shift register. In $1/n$-rate codes, a single bit is fed into the shift register at a time, thus there are two possible state transitions. This results in a trellis diagram where the transition form a perfect shuffle as depicted in Fig. 4a. In $2/n$-rate codes, two bits enter the shift register at a time, thus four state transitions are possible, which results in a stride-by-4 permutation in the interconnection as shown in Fig. 4b.

The previous examples show that stride permutations have practical and important applications in the fields of digital signal processing and telecommunications. Applications in these areas are often hard real-time constrained and realized in systems with relatively low clock frequencies (e.g., for extending battery life). Therefore, parallel implementations are preferred, which implies also the need to access several operands simultaneously to increase the memory bandwidth.

Typical realization for all the previous applications are recursive, i.e., small kernels operate over a data array and the results of an iteration are used as operands in the next iteration. In addition, read operations are performed in different order than write operations; e.g., in the Viterbi decoding of code illustrated in Fig. 4a, operands are read in perfect shuffle order ($P_{16,8}$) and the

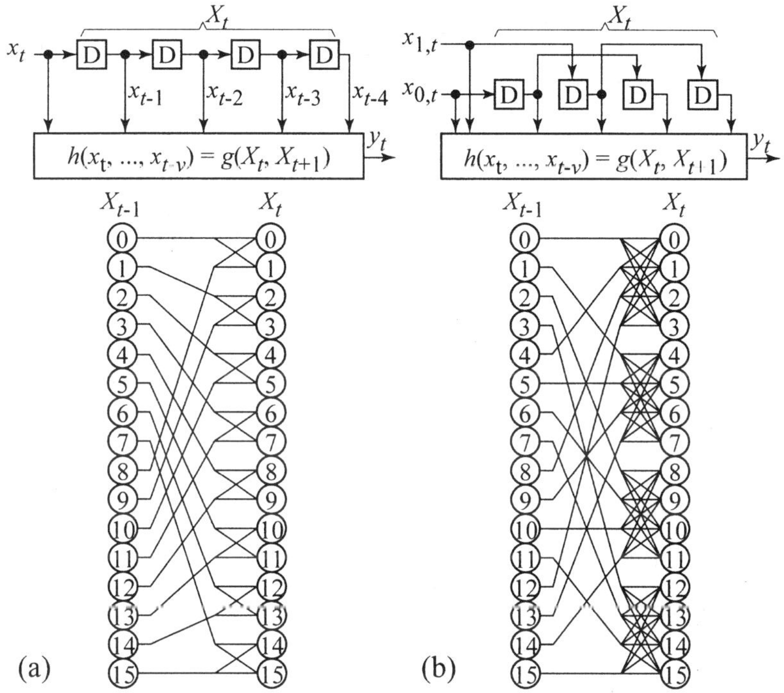

Figure 4 Single-shift register convolutional encoders and allowed state transitions: (a) $1/n$-rate code and (b) $2/n$-rate code. (x_t: input at time instant t. X_t: state at time instant t. y_t: output at time instant t. D: bit register.)

results are stored in linear order ($P_{16,1}$). In order to minimize memory consumption, the results should be stored into the same memory locations where the operands were obtained. However, after the first iteration the results will be in perfect shuffle order $P_{N,N/2}$, not in linear order, $P_{N,1}$, as intended originally. According to Corollary 1, the next read access should be performed in $P_{N,N/2}$ order to compensate the previous additional reordering. Respectively, the next read should be according to $P_{N,N/8}$. Eventually we find that $\log_2 N$ different strides are needed (i.e., all the strides of power-of-two from 1 to $N/2$).

The need for an access scheme for stride permutations can be illustrated with an example by referring to Fig. 2. In this example, a 32-element array (0, 1, . . . , 31) is distributed over four memory modules. The possible stride permutation accesses in this case are $P_{32,1}$, $P_{32,2}$, $P_{32,4}$, $P_{32,8}$, and $P_{32,16}$. The low-order interleaving in Fig. 2a allows only conflict-free access for linear access, $P_{32,1}$, and all the others introduce conflicts. The row rotation and linear transformation schemes in Fig. 2b and 2c, respectively, provide conflict-free access for $P_{32,1}$, $P_{32,2}$, and $P_{32,4}$. By noting that the elements 0,

8, and 16 are stored into the same module, we find that accesses $P_{32,8}$ and $P_{32,16}$ introduce conflicts. Especially the perfect shuffle access $P_{32,16}$ is difficult: the accesses should be performed in the following order: ([0, 16, 1, 17], [2, 18, 3, 19], [4, 20, 5, 21], [6, 22, 7, 23], [8, 24, 9, 25], [10, 26, 11, 27], [12, 28, 13, 29], [14, 30, 15, 31]). The linear transformation scheme in Fig. 2d has conflict only in this access pattern.

In the previously reported access schemes, stride access has been defined to access every kth element, while in stride permutation access the pattern wraps into the beginning of the array. Especially in perfect shuffle access $P_{N,N/2}$, elements with distances of 1 and $N/2$ need to be accessed, which is not supported by stride access. This illustrates the principal difference between the stride access and stride permutation access.

V. CONFLICT-FREE PARALLEL MEMORY ACCESS FOR STRIDE PERMUTATION

The previous discussion shows that there is a need to perform several memory accesses in parallel. In particular, it was found that when an N-element data array is accessed according to a stride permutation, there will be a need to support several strides. Often the array lengths are powers-of-two and in such cases all the power-of-two strides from 1 to $N/2$ are needed. Furthermore, in practical systems, the numbers of system elements (e.g., processing elements or memory modules), is often a power-of-two.

In this section, we present a conflict-free parallel access scheme supporting various stride permutation accesses to a data array in matched memory systems. The proposed method is based on linear transformations since they suit systems better with a power-of-two number of memory modules. Although Theorem 3 suggests that a conflict-free access scheme supporting multiple strides cannot be designed, we may, however, relax the constraints.

A. Assumptions

In order to derive a stride permutation access scheme, we make the following assumptions; (a) the array length is constant and power-of-two, $N = 2^n$; (b) the array is stored in n-word boundaries; (c) the number of memory modules is a power-of-two, $Q = 2^q$; and (d) the strides in stride permutation access are powers-of-two, $S = 2^s$.

Assumption (a) implies that constraints on the initial address need to be set resulting in assumption (b). Such a constraint has already been used in several commercial DSP processors for performing circular addressing [20]. Assumption (c) is actually a practical assumption in digital systems. Assumption (d) implies that the address mapping should produce a q-bit memory module address and an $(n - q)$-bit row address. All these assumptions may be considered practical.

B. Access Scheme

We may develop an access scheme for stride permutation by using the previously discussed principle for generating the row address; according to Eq. (7), row address $r = (r_{n-q-1}, r_{n-q-2}, \ldots, r_0)^T$ is obtained by extracting the $(n - q)$ most significant bits from the address:

$$r_i = a_{i+q}, i = 0, 1, \ldots, n - q - 1 \tag{16}$$

The previous assumptions define that transformation matrices will be specific for each array length N and number of modules Q. However, the stride is no longer a parameter for the matrix. Therefore, we introduce a new notation for the linear transformation matrix: $T_{N,Q}$, which defines clearly the array length and number of modules.

The discussion in Section 2 related to the example in Fig. 2 implies that the periodicity of the linear transformation scheme used in the example is not large enough. This can be clearly seen by comparing the order of elements in each row; the ordering repeats after the fourth column (i.e., the period is 16). This is already reflected by the fact that the module address is generated by using four bits from the address.

The periodicity can be increased by adding the number of bits affecting the module address. This was already suggested in [15] for unmatched memory systems but the additional bit fields are only copied, not included into the bit-wise XOR operations. A special case of perfect shuffle access is discussed in [21], where two elements are accessed from a two-memory system, $Q = 2$. In such a case, the module address is defined by the parity of the address, thus the transformation matrix $T_{2^n,2}$ is a vector of n elements of 1s. This implies that the additional bits should be included into the bit-wise XOR operations (i.e., each row in $T_{N,Q}$ should contain multiple 1s).

The use of diagonals were suggested in [13], thus the obvious solution would be to add diagonals to T. Let us illustrate this approach with an example of a 64-element array is distributed over four memory modules. In such a case, the transformation matrix $T_{64,4}$ would be the following:

$$T_{64,4} = \begin{pmatrix} 1 & 0 & 1 & 0 & 1 & 0 \\ 0 & 1 & 0 & 1 & 0 & 1 \end{pmatrix} \tag{17}$$

This will result in the storage depicted in Fig. 5 and it is easy to see that all the stride permutation accesses with power-of-two strides from 1 to 32 are conflict-free. Performed computer simulations verified that the transforma-

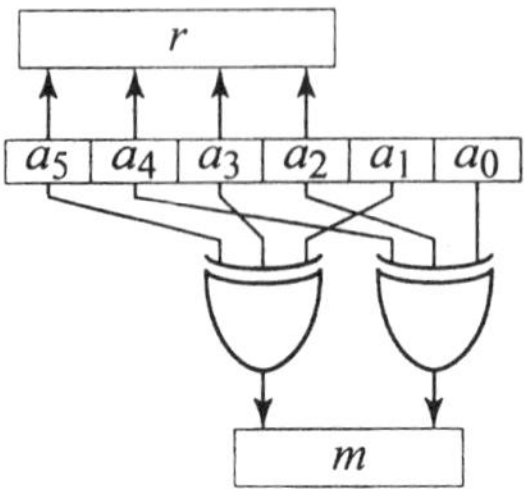

(a)

(b)

m	0	1	2	3
	0	1	2	3
	5	4	7	6
	10	11	8	9
	15	14	13	12
	17	16	19	18
	20	21	22	23
	27	26	25	24
	30	31	28	29
	34	35	32	33
	39	38	37	36
	40	41	42	43
	45	44	47	46
	51	50	49	48
	54	55	52	53
	57	56	59	58
	60	61	62	63

Figure 5 Access scheme for 64-element array on a 4-module system corresponding to the transformation matrix in Eq. (17): (a) module address generation and (b) contents of memory modules.

tion matrix can be designed by filling the matrix with $q \times q$ diagonals in cases where n rem $q = 0$; transformation matrices $T_{k2^q,2^q}$ can be obtained by concatenating k identity matrices I_q.

The next question is how the matrix is formed when n rem $q \neq 0$. For this purpose, additional 1s need to be included into $T_{N,Q}$. In [13], such 1s were added as diagonals or antidiagonals off the main diagonals. In the approach proposed in [12], the main diagonals may contain 0s thus the additional 1s are spread over the matrix to fulfil the full rank requirement. This results in the fact that rows may contain large number of ones, thus the number of bits needed in XOR-operations is increased. The effect is even worse in the approach used in [14], where the rows may contain different numbers of 1s; one row is full of 1s, another contains only a single 1. This is extremely uncomfortable from the implementation point of view when several array lengths need to be supported since the transformation matrix will be different for different array lengths. In such a case, the number of bits that have XOR together varies from 1 to n.

The previous discussion implies that the additional bits should be concentrated to the right part of $T_{N,Q}$ (i.e., to A_L in the original matrix T in Eq.(8).) Such an arrangement eases the configuration of the address generation when the array length changes. If the 1s are in matrix A_H, the address bits a_i, which need to be included into XOR operations may change and require multiplexing. Now, if all the configurations are performed for the least significant bits of a, these are always in the same position independent of the array length.

Therefore, in cases where n rem $q \neq 0$, we fill the transformation matrix with diagonals starting from the right lower corner and, if there is not enough space available in the left of the matrix, a partial diagonal is placed. The remaining partial diagonal wraps back to the right and filling begins from the rightmost column of $T_{N,Q}$ in a row, which is above the row where the last 1 in the leftmost column was placed. If the diagonal will hit the top row, it will be continued from the bottom row in the preceding column. A total of $(n + q - \gcd(q, n \bmod q))$ ones will be used in $T_{2^n,2^q}$, where gcd(·) is the greatest common denominator. The entire access scheme can be formalized as follows:

$$m_i = \bigoplus_{k=0}^{l_{n,q}(i)} a_{(jq+i) \bmod n}, \qquad i = 0, 1, \ldots, q-1$$
$$l_{n,q}(i) = \lfloor (n + q - \gcd(q,\ n \bmod q) - i - 1)/q \rfloor \tag{18}$$

where $\oplus$ denotes the bit-wise XOR operation. The row address is obtained according to Eq. (16).

This approach provides a solution to the example shown in Fig. 2 and the transformation matrix $T_{32,4}$ is

$$T_{32,4} = \begin{pmatrix} 0 & 1 & 0 & 1 & 1 \\ 1 & 0 & 1 & 0 & 1 \end{pmatrix} \tag{19}$$

The contents of the memory modules stored according to $T_{32,4}$ is illustrated in Fig. 6. Once again it can be seen that the stride permutation access is supported for all the strides of power-of-two from 1 to 16.

(a)

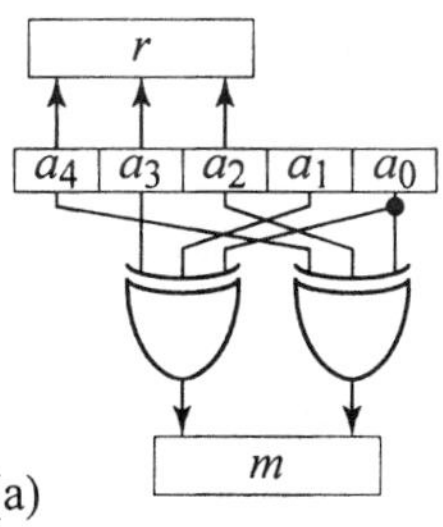

(b)

m	0	1	2	3
	0	3	2	1
	7	4	5	6
	10	9	8	11
	13	14	15	12
	19	16	17	18
	20	23	22	21
	25	26	27	24
	30	29	28	31

Figure 6 Access scheme for 32-element array on a 4-module system corresponding to the transformation matrix in Eq. (19): (a) module address generation and (b) contents of memory modules.

C. Validation

The presented access scheme was verified with computer simulations by generating storage organizations and verifying that each access is conflict-free. For a given array length $N = 2^n$, the number of memory modules Q was varied to cover all the possible numbers of power-of-two (i.e., $Q = 2^0, 2^1, \ldots, 2^{n-1}$). For each parameter pair (N,Q), all the stride permutation accesses were performed with strides covering all the possible powers-of-two: $S = 2^0, 2^1, \ldots, 2^{n-1}$ and each parallel access was verified to be conflict-free. The power-of-two array lengths were iterated from 2^1 to 2^{20}. The extensive simulation did not find any conflicts and, therefore, we can state that the presented access scheme provides conflict-free parallel stride permutation access in practical cases (i.e., array lengths up to 2^{20}), for all the possible power-of-two strides on matched interleaved memory systems where the number of memory modules is a power-of-two.

D. Address Generation

Before going into implementations, we may investigate the structure $T_{N,Q}$ when N is varied. In practical systems, Q is constant; the number of memory modules is only a design time parameter. For example, module transformation matrices for 16- and 64-module systems are illustrated in Fig. 7 and a few observations can be made from the structure of these matrices.

First, the matrices contain two principal diagonal structures: concatenated diagonals from the bottom-right corner to left and additional off-diagonals. The concatenated diagonals imply that the address a should be divided into q-bit fields and bit-wise XOR is performed between these fields. Since the concatenated diagonals in matrix for a shorter array is included in a matrix for longer arrays, several array lengths can be supported easily; shorter arrays can be supported by feeding 0s to the most significant address bits.

The second observation is that the off-diagonals affect at most the $q - 1$ least significant bits of address a. In fact, Eq. (18) dictates that the number of 1s in off-diagonals is $q - \gcd(q, n \bmod q)$. In addition, the structure of off-diagonals depends on the relation between n and q but, since, in practice, q is constant, the structure depends on array length. However, there are only q different structures; the off-diagonal structure has periodic behavior when the array length is increasing. In Fig. 7, one complete period is shown and $T_{8192,64}$ would have the same off-diagonal structure as $T_{128,64}$.

The structure of off-diagonals implies that several array lengths can be supported if a predetermined control word configures additional hardware to perform the functionality of the off-diagonals. Such a configuration is

$$T_{32,16} = \begin{pmatrix} 0&1&1&0&0\\ 0&0&1&1&0\\ 0&0&0&1&1\\ 1&0&0&0&1 \end{pmatrix} \qquad T_{128,64} = \begin{pmatrix} 0&1&1&0&0&0&0\\ 0&0&1&1&0&0&0\\ 0&0&0&1&1&0&0\\ 0&0&0&0&1&1&0\\ 0&0&0&0&0&1&1\\ 1&0&0&0&0&0&1 \end{pmatrix}$$

$$T_{64,16} = \begin{pmatrix} 0&0&1&0&1&0\\ 0&0&0&1&0&1\\ 1&0&0&0&1&0\\ 0&1&0&0&0&1 \end{pmatrix} \qquad T_{256,64} = \begin{pmatrix} 0&0&1&0&1&0&0&0\\ 0&0&0&1&0&1&0&0\\ 0&0&0&0&1&0&1&0\\ 0&0&0&0&0&1&0&1\\ 1&0&0&0&0&0&1&0\\ 0&1&0&0&0&0&0&1 \end{pmatrix}$$

$$T_{128,16} = \begin{pmatrix} 0&0&0&1&0&0&1\\ 1&0&0&0&1&0&0\\ 0&1&0&0&1&1&0\\ 0&0&1&0&0&1&1 \end{pmatrix} \qquad T_{512,64} = \begin{pmatrix} 0&0&0&1&0&0&1&0&0\\ 0&0&0&0&1&0&0&1&0\\ 0&0&0&0&0&1&0&0&1\\ 1&0&0&0&0&0&1&0&0\\ 0&1&0&0&0&0&0&1&0\\ 0&0&1&0&0&0&0&0&1 \end{pmatrix}$$

$$T_{256,16} = \begin{pmatrix} 1&0&0&0&1&0&0&0\\ 0&1&0&0&0&1&0&0\\ 0&0&1&0&0&0&1&0\\ 0&0&0&1&0&0&0&1 \end{pmatrix} \qquad T_{1024,64} = \begin{pmatrix} 0&0&0&0&1&0&0&0&1&0\\ 0&0&0&0&0&1&0&0&0&1\\ 1&0&0&0&0&0&1&0&0&0\\ 0&1&0&0&0&0&0&1&0&0\\ 0&0&1&0&0&0&1&0&1&0\\ 0&0&0&1&0&0&0&1&0&1 \end{pmatrix}$$

$$T_{512,16} = \begin{pmatrix} 0&1&0&0&0&1&1&0&0\\ 0&0&1&0&0&0&1&1&0\\ 0&0&0&1&0&0&0&1&1\\ 1&0&0&0&1&0&0&0&1 \end{pmatrix} \qquad T_{2048,64} = \begin{pmatrix} 0&0&0&0&0&1&0&0&0&0&1\\ 1&0&0&0&0&0&1&0&0&0&0\\ 0&1&0&0&0&0&1&1&0&0&0\\ 0&0&1&0&0&0&0&1&1&0&0\\ 0&0&0&1&0&0&0&0&1&1&0\\ 0&0&0&0&1&0&0&0&0&1&1 \end{pmatrix}$$

$$T_{1024,16} = \begin{pmatrix} 0&0&1&0&0&0&1&0&1&0\\ 0&0&0&1&0&0&0&1&0&1\\ 1&0&0&0&1&0&0&0&1&0\\ 0&1&0&0&0&1&0&0&0&1 \end{pmatrix} \qquad T_{4096,64} = \begin{pmatrix} 1&0&0&0&0&0&1&0&0&0&0&0\\ 0&1&0&0&0&0&0&1&0&0&0&0\\ 0&0&1&0&0&0&0&0&1&0&0&0\\ 0&0&0&1&0&0&0&0&0&1&0&0\\ 0&0&0&0&1&0&0&0&0&0&1&0\\ 0&0&0&0&0&1&0&0&0&0&0&1 \end{pmatrix}$$

(a) (b)

Figure 7 Transformation matrices for module address generation in (a) 16- and (b) 64-module systems.

actually simple by noting that the form of off-diagonals in different array lengths indicates rotation of least significant bits in a. The number of bits rotated is dependent on the relation between n and q.

According to the previous observations, the computation of the module address m can be interpreted as follows. First, the address a is divided into q-bit fields, F^i, starting from the least significant bit of a, i.e., $F^i = (a_{iq+q-1}, a_{iq+q-2}, \ldots, a_{iq+1}, a_{iq})^T$. If $e = n \bmod q > 0$, the most significant bits e of a exceeding the q-bit block border are extracted as a bit vector L, i.e.,

$$L = (a_{n-1}, a_{n-2}, \dots, a_{n-e})^T \tag{20}$$

Next, a q-bit field $X = (x_{q\text{-}1}, x_{q\text{-}2}, \dots, x_0)$ is formed by extracting the $(q - \gcd(q,e))$ least significant bits of the address a and placing zeroes to the most significant bits;

$$X = (0, \dots, 0, a_{q-\gcd(q,e)} - 1, \dots, a_1, a_0)^T \tag{21}$$

The bit vector X is rotated $g = (n - q \bmod q)$ bits to the left to obtain a bit vector $O = (o_{q-1}, \dots, o_0)^T$, i.e.,

$$O = \mathrm{rot}_{(n-q) \bmod q}(X) \tag{22}$$

where $\mathrm{rot}_g(\cdot)$ denotes g-bit left rotation (circular shift) of the given bit vector, i.e.,

$$\mathrm{rot}_g((a_{k-1}, a_{k-2}, \dots, a_0)^T) = \\ (a_{k-g-1}, a_{k-g-2}, \dots, a_0, a_{k-1}, \dots, a_{k-g+1}, a_{k-g})^T \tag{23}$$

Finally the module address m is obtained by performing bit-wise XOR operation between the vectors F_i, X, and L:

$$m_i = \begin{cases} o_i \oplus (\oplus_{j=0}^{\lfloor n/q \rfloor - 1} a_{jq+i}), & i \geq e \\ l_i \oplus o_i \oplus (\oplus_{j=0}^{\lfloor n/q \rfloor - 1} a_{jq+i}), & i < e \end{cases} \tag{24}$$

A principal block diagram of the module address generation according to the previous interpretation is illustrated in Fig. 8. This block diagram contains a rotation unit shown in Fig. 9, which computes the vector O. This unit obtains $q - 1$ least significant bits of a as an input and the $\gcd(q,e) - 1$ most significant bits of input are zeroed, thus the $q - \gcd(q,e)$ least significant bits are passed through to form a q-bit vector $X = (0, \dots, 0, a_{q\text{-}\gcd}(q,e){-}1, a_{q-\gcd}(q,e){-}2, \dots, a_1, a_0)^T$. These bits can be selected with the aid of a bit vector $f = (f_{g-2}, \dots, f_1, f_0))^T$, where the $q - \gcd(q,e)$ least significant bits are 1s and the $\gcd(q,e) - 1$ most significant bits are 0s. A bit-wise AND operation is performed with the input vector and the obtained vector X is then rotated g bits to the left and the q-bit vector O is obtained.

The main advantage of this scheme can be seen from the block diagram in Fig. 8. In the address generation, each individual XOR is performed on, at most, $(\lfloor n/q \rfloor + 2)$ bit lines while in other schemes, e.g., in [14], some XORs require all the n address bits. which complicates implementation when several array lengths need to be supported.

The support for different array lengths in the implementation requires only a single predetermined control word defining the bit selection and ro-

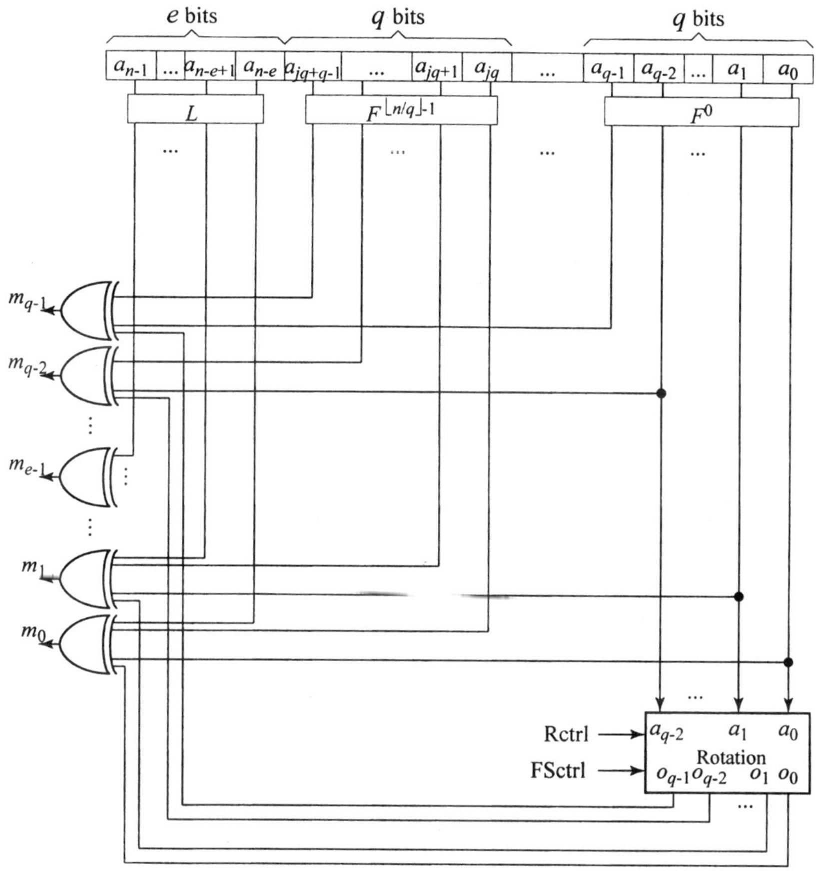

Figure 8 Principal block diagram of module address generation. (Rctrl: rotation control. FSctrl: field selection control.)

tation. This control word needs to be modified only when the length of the array to be accessed is changed. There is no need to store the complete transformation matrix as in some proposed realizations, e.g., in [14].

VI. SUMMARY

In this chapter, we briefly reviewed the principal access schemes for interleaved memory systems for increasing the data transfer rate between memory

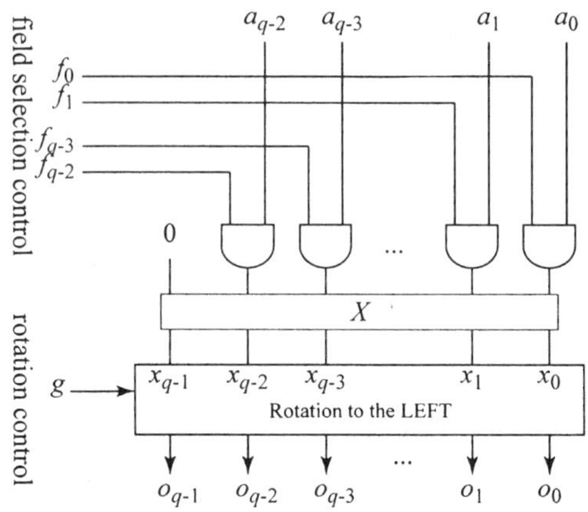

Figure 9 Principal block diagram of rotation unit in module address generation.

and computational resources. Such schemes were studied extensively in the supercomputer area but only a few studies considered the technique of embedded systems. An introduction to stride access was given and it was found that a conflict-free access scheme supporting several strides is not possible without including initial constraints. We defined stride permutation access and gave examples of its applications. It was shown that stride permutations are found in several DSP applications where small kernels are iterated, thus a special addressing scheme supporting the access pattern provides advantage especially when long arrays are used.

In this chapter, a conflict-free stride permutation access scheme for matched memory systems was presented. It was assumed that 2^n data elements are distributed over 2^q independent memory modules. The used assumptions dictate that the array length is constant and the initial address is zero. In this case, all the possible power-of-two stride permutation accesses are conflict-free, which was verified with computer simulations. The module address generation is simple and requires only bit-wise XOR operations.

It was shown that several array lengths can be supported by including a q-bit left shifter into the module address generator. In this case, all additional operations are performed on the $q - 1$ least significant bits of the address independent on the array length. The presented scheme can support different initial addresses but arrays need to be stored into n-word boundaries. However, this is not a strict requirement and such a restriction is already present in some addressing modes in commercial DSP processors. This access scheme can be used in application-specific array processors where operands need to be reordered according to stride permutation. In such cases, multiported memories or double buffering can be avoided when an interleaved memory system is used.

REFERENCES

1. Seznec A., Lenfant J. (1994). Interleaved parallel schemes. *IEEE Trans. Parallel and Distrib. Syst.* 5(12):1329–1334.
2. Wuytack S., Catthoor F., de Jong G., De Man H. J. (1999). Minimizing the required memory bandwidth in VLSI system realizations. *IEEE Trans. VLSI Syst.* 7(4):433–441.
3. Lin H., Wolf W. (2000). Co-design of interleaved memory systems. *Proc. Int. Workshop Hardware/Software Codesign, San Diego, CA, 46–50.*
4. Chen S., Postula A. (2000). Synthesis of custom interleaved memory systems. *IEEE Trans. VLSI Syst.* 8(1):74–83.
5. Harper D. T. Increased memory performance during vector accesses through the use of linear address transformations. *IEEE Trans. Comput.* 41(2):227–230.
6. Budnik P., Kuck D. (1971). The organization and use of parallel memories. *IEEE Trans. Comput.* 20(12):1566–1569.
7. Gössel, M., Rebel, B., Creutzburg, R. (1994). Memory Architecture & Parallel Access. Amsterdam, The Netherlands: North Holland.
8. Lawrie D.H., Vora C.R. (1982). The prime memory system for array access. *IEEE Trans. Comput.* 31(5):435–442.
9. Wijshoff H.A.G., van Leeuwen J. (1985). The structure of periodic storage schemes for parallel memories. *IEEE Trans. Comput.* 34(6):501–505.
10. Deb A. (1996). Multiskewing—a novel technique for optimal parallel memory access. *IEEE Trans. Parallel and Distrib. Syst.* 7(6):595–604.
11. Frailong J.M., Jalby W., Leflant J. (1985). XOR-schemes: a flexible data organization in parallel memories. *Proc. Int. Conf. Parallel Processing, St. Charles, IL,* 276–283.
12. Sohi G.S. (1993). Interleaved memories for vector processors. *IEEE Trans. Comput.* 42(1):34–44.
13. Harper D.T. Block, multistride vector, and FFT accesses in parallel memory systems. *IEEE Trans. Parallel and Distrib. Syst.* 2(1):43–51.
14. Norton A., Melton E. (1987). A class of boolean linear transformations for conflict-free power-of-two stride access. *Proc. Int. Conf. Parallel Processing, St. Charles, IL, 247–254.*
15. Valero M., Lang T., Peiron M., Ayguadé E. (1995). Conflict-free access for streams in multimodule memories. *IEEE Trans. Comput.* 44(5):634–646.
16. Granata J., Cooner M., Tolimieri R. (1992). Recursive fast algorithms and the role of the tensor product. *IEEE Trans. Signal Processing* 40(12):2921–2930.
17. Davio M. (1981). Kronecker products and shuffle algebra. *IEEE Trans. Comput.* 30(2):116–125.

18. Astola J., Akopian D. (1999). Architecture-oriented regular algorithms for discrete sine and cosine transforms. *IEEE Trans. Signal Processing* 47(4):1109–1124.
19. Takala J., Akopian D., Astola J., Saarinen J. (2000). Constant geometry algorithm for discrete cosine transform. *IEEE Trans. Signal Processing* 48(6): 1840–1843.
20. Lapsley P.D., Bier J., Shoham A., Lee E.A. (1996). DSP Processor Fundamentals: Architectures and Features. Fremont, CA: Berkeley Design Technology, Inc.
21. Cohen D. (1976). Simplified control of FFT hardware. *IEEE Trans. Acoust., Speech, Signal Processing* 24(6):255–579.

5

On Modeling Intra-Task Parallelism in Task-Level Parallel Embedded Systems

Andy D. Pimentel, Frank P. Terpstra, Simon Polstra, and Joe E. Coffland

University of Amsterdam, Amsterdam, The Netherlands

I. INTRODUCTION

Modern embedded systems, like those for media and signal processing, often have a heterogeneous system architecture, consisting of components in the range from fully programmable processor cores to dedicated hardware components. Increasingly, these components are integrated as a system-on-chip exploiting task-level parallelism in applications. Due to the high degree of programmability that is usually provided by such embedded systems, they typically allow for targeting a whole range of applications with varying demands. All of the above characteristics greatly complicate the design of these embedded systems, making it more and more important to have good tools available for exploring different design choices at an early stage in the design.

In the context of the Artemis project (ARchitectures and meThods for Embedded MedIa Systems) [20], we are developing an architecture workbench that provides modeling and simulation methods and tools for the efficient design space exploration of heterogeneous embedded multimedia systems. This architecture workbench should allow for rapid performance evaluation of different architecture designs, application to architecture mappings, and hardware/software partitionings and it should do so at multiple levels of abstraction *and* for a wide range of multimedia applications.

In this chapter, our focus is on a prototype modeling and simulation environment called Sesame [19]. According to the Artemis modeling methodology [20], this environment uses separate application models and architecture models, and an explicit mapping step to map an application model onto an architecture model. This mapping is realized by means of trace-driven co-simulation, where the execution of the application model generates application events that represent the application workload imposed on the architecture. Application models consist of communicating parallel processes, thereby expressing the task-level parallelism available in the applications. By mapping the event traces generated by different application model processes onto the various system architecture components, this task-level parallelism is exploited at the architectural level. In addition, the underlying architecture may also exploit intra-task parallelism inside a single trace. This chapter presents the newly added techniques Sesame applies to model architectures that exploit such intra-task parallelism. Moreover, using a case study with the QR decomposition algorithm as application, we demonstrate the effectiveness of our modeling methodology.

The remainder of this chapter is organized as follows. Section II briefly describes related work in the area of modeling and simulation of complex embedded systems. Section III gives a general overview of the Sesame modeling and simulation environment, while in Section IV we present a more detailed description of Sesame's synchronization layer. In Sections V and VI, we describe the methods applied to model intra-task parallelism and discuss their impact on Sesame's synchronization and architecture model layers. Section VII presents some validation results we obtained from the case study with the QR decomposition application. Finally, Section VIII discusses several open issues and Section IX concludes the chapter.

II. RELATED WORK

Various research groups are active in the field of modeling and simulating heterogeneous embedded systems, of which some are academic efforts (e.g., [6,12,10]) and others commercial [9] and industrial efforts (e.g., [5]). Many efforts in this field *co-simulate* the software parts, which are mapped onto a programmable processor, with the hardware components and their interactions together in one simulation. Because an explicit distinction is made between software and hardware simulation, it must be known which application components will be performed in software and which ones in hardware before a system model is built. This significantly complicates the performance evaluation of different hardware/software partitioning schemes since a new system model may be required for the assessment of each partitioning.

A number of exploration environments, such as VCC [1], Polis [4] and eArchitect [2], facilitate more flexible system-level design space exploration by providing support for mapping a behavioral application specification to an architecture specification. Within the Artemis project, however, we try to push the separation of modeling application behavior and modeling architectural constraints at the system level to even greater extents. To this end, we apply trace-driven co-simulation of application and architecture models. As was shown in [19], this leads to efficient exploration of different design alternatives while also yielding a high degree of reusability. The work of [16] also used a trace-driven approach, but this was done to extract communication behavior for studying on-chip communication architectures. Rather than using the traces as input to an architecture simulator, their traces were analyzed statically. In addition, a traditional hardware/software co-simulation stage is required in order to generate the traces.

Finally, the Archer project [23] shows a lot of similarities with our work. This is due to the fact that both our work and Archer are spin-offs from the Spade project [18]. A major difference is, however, that Archer follows an entirely different application-to-architecture mapping approach. Instead of using event-traces, it maps symbolic programs, which are derived from the application model, onto architecture model comonents.

III. THE SESAME MODELING AND SIMULATION ENVIRONMENT

The Sesame modeling and simulation environment [19], which builds upon the ground-laying work of the Spade framework [18], facilitates the performance analysis of embedded systems architectures in a way that directly reflects the so-called Y-chart design approach [14]. In Y-chart based design, a designer studies the target applications, makes some initial calculations, and proposes an architecture. The performance of this architecture is then quantitatively evaluated and compared against alternative architectures. For such performance analysis, each application is mapped onto the architecture under investigation and the performance of each application–architecture combination is evaluated. Subsequently, the resulting performance numbers may inspire the designer to improve the architecture, restructure the application(s), or modify the mapping of the application(s).

In accordance to the Y-chart approach, Sesame recognizes separate application and architecture models within a system simulation. An application model describes the functional behavior of an application, including both computation and communication behavior. The architecture model defines architecture resources and captures their performance constraints. Essential in this modeling methodology is that an application model is independent from architectural specifics, assumptions on hardware/software partitioning, and timing characteristics. As a result, a single application model can be used to exercise different hardware/software partitionings and can be mapped onto a range of architecture models, possibly representing different system architectures or simply modeling the same system architecture at various levels of abstraction. After mapping, an application model is co-simulated with an architecture model allowing for evaluation of the system performance of a particular application, mapping, and underlying architecture.

For application modeling, Sesame uses the Kahn process network (KPN) model of computation [13] in which parallel processes—implemented in a high-level language—communicate with each other via unbounded FIFO channels. In the Kahn paradigm, reading from channels is done in a blocking manner, while writing is nonblocking. The computational behavior of an application is captured by instrumenting the code of each Kahn process with annotations which describe the application's computational actions. The reading from or writing to Kahn channels represents the communication behavior of a process within the application model. By executing the Kahn model, each process records its actions in order to generate a trace of application events, which is necessary for driving an architecture model. Initially, the application events typically are coarse-grained, such as *execute(DCT)* or *read(pixel-block, channel_id)*, and they may be refined as the underlying architecture models are refined. We note that in the remainder of this chapter, computational application events will be referred to as *execute events*.

To execute Kahn application models, and thereby generating the application events that represent the workload imposed on the architecture, Sesame features a process network execution engine supporting Kahn semantics. This execution engine runs the Kahn processes as separate threads using the Pthreads package. For now, there is a limitation that the Kahn processes need to be written in C++. In the near future, C and Java support will be added. The structure of the application models (i.e., which processes are used in the model and how they are connected to each other) is described in a language called Y-chart Modeling Language (YML) [8]. This is an XML-based language that is similar to Ptolemy's MoML [17] but is slightly less generic in the sense that YML only needs to support a few simulation domains. As a consequence, YML only supports a subset of MoML's features. However, YML provides one additional feature in comparison to MoML as it contains built-in scripting support. This allows for loop-like constructs, mapping and connectivity functions, and so on, which facilitate the description of large and complex models.

The performance of an architecture can be evaluated by simulating the performance consequences of the incoming execute and communication

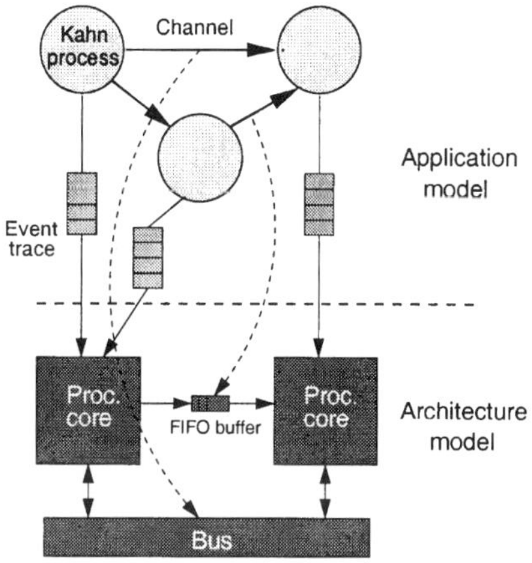

Figure 1 Mapping a Kahn application model onto an architecture model.

events from an application model. This requires an explicit mapping of the processes and channels of a Kahn application model onto the components of the architecture model. The generated trace of application events from a specific Kahn process is therefore routed toward a specific component inside the architecture model by using a trace-event queue. This is illustrated in Fig. 1. Since the application-model execution engine and the architecture simulator run as separate processes,* these trace-event queues are currently implemented via Unix named-pipes. Alternative implementations of the queues, such as using shared memory, are envisioned in the future. If two or more Kahn processes are mapped onto a single architecture component (e.g., when several application tasks are mapped onto a microprocessor), then the events from the different trace-event queues need to be scheduled. The next section explains how this is done.

An architecture model solely accounts for architectural (performance) constraints and therefore does not need to model functional behavior. This is possible because the functional behavior is already captured in the application model, which subsequently drives the architecture simulation. An architecture model is constructed from generic building blocks provided by a library.

* Running the application-model execution engine as a separate process also makes it easy to analyze the application model in isolation. This can be beneficial as it allows for investigation of the upper bounds of the performance and may lead to early recognition of bottlenecks within the application itself.

This library contains template performance models for processing cores, communication media (like busses) and different types of memory. These template models can be freely extended and adapted. All architecture models in Sesame are implemented using a small but powerful discrete-event simulation language, called Pearl, which provides easy construction of the models and fast simulation [19]. The structure of architecture models—specifying which building blocks are used from the library and the way they are connected—is also described in YML.

IV. THE SYNCHRONIZATION LAYER

When multiple Kahn application model processes are mapped onto a single architecture model component, the event traces need to be scheduled. For this purpose, Sesame provides an intermediate *synchronization layer*, which is illustrated in Fig. 2. This layer guarantees deadlock-free scheduling of the application events and forms the application and architecture dependent structure that connects the architecture-independent application model with the application-independent architecture model. The synchronization layer,

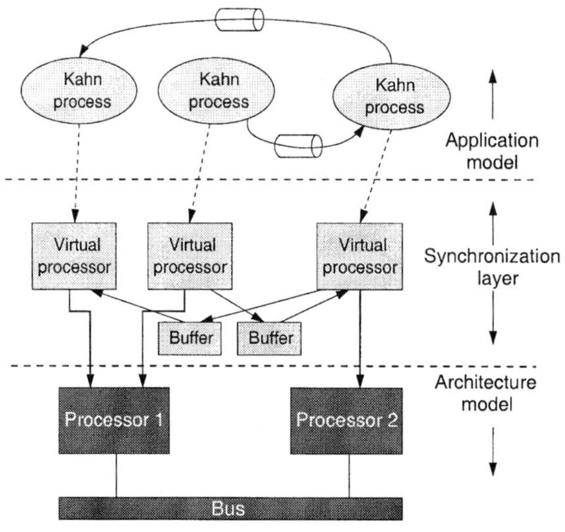

Figure 2 The three layers within Sesame: the application model layer, the architecture model layer, and the synchronization layer which interfaces between application and architecture models.

which can be automatically generated from the YML description of an application model, consists of virtual processor components and FIFO buffers for communication between the virtual processors. There is a one-to-one relationship between the Kahn processes in the application model and the virtual processors in the synchronization layer. This is also true for the Kahn channels and the FIFO channels in the synchronization layer, except for the fact that the buffers of the latter channels are limited in size. Their size is parameterized and dependent on the modeled architecture. A virtual processor reads in an application trace from a Kahn process and dispatches the events to a processing component in the architecture model. The mapping of a virtual processor onto a processing component in the architecture model is parameterized and thus freely adjustable. Currently, this virtual processor to architectural processor mapping is specified in the YML description of the architecture model. We are working, however, toward an approach in which this mapping is specified in a separate YML mapping description.

As can be seen from Fig. 2, multiple virtual processors can be mapped onto a single processor in the architecture model. In this scheme, execute events are directly dispatched by a virtual processor to the processor model. The latter subsequently schedules the events originating from different virtual processors according to some given policy (FCFS by default) and models their timing consequences. For communication events, however, the appropriate buffer at the synchronization layer is first consulted to check whether or not a communication is safe to take place so that no deadlock can occur. Only if it is found to be safe (i.e., for read events the data should be available and for write events there should be room in the target buffer), then communication events may be dispatched to the processor component in the architecture model. As long as a communication event cannot be dispatched, the virtual processor blocks. This is possible because the synchronization layer is, like the architecture model, implemented in the Pearl simulation language and executes in the same simulation-time domain as the architecture model. As a consequence, the synchronization layer accounts for synchronization delays of communicating application processes mapped onto the underlying architecture, while the architecture model accounts for the computational latencies and the pure communication latencies (e.g., bus arbitration and transfer latencies). Each time a virtual processor dispatches an application event (either computation or communication) to a processor in the architecture model, it is blocked in simulated time until the event's simulation at the architecture level has finished.

The idea of concentrating synchronization behavior in a synchronization layer and separating it from (the latencies caused by) data transmission behavior is somewhat similar to the synchronization graph concept of [21].

However, our synchronization layer seems to be more flexible since it is dynamically scheduled and behaves like a "Kahn" process network in which the FIFO buffers are bounded. As a consequence of the dynamic scheduling of the synchronization layer and the architecture model (remember that they both are executed in the same discrete-event simulation domain), dynamics at the architecture level such as contention can easily be taken into account within the synchronization layer.

V. MODELING INTRA-TASK PARALLELISM

Initially, Sesame only modeled the architecture's processing cores as black boxes which sequentially simulate the timing consequences of the incoming (linear) trace of application events. However, the architecture under investigation may also want to exploit intra-task parallelism which is present in a single event trace from a Kahn application process. For example, a processing element may have multiple communication units that perform independent reads and writes in parallel, or it may have multiple execution units for concurrently processing independent computations. To support the modeling and simulation of such intra-task parallelism, we extended Sesame's model library with component models that allow for refining the virtual processors in the synchronization layer and the processor components within the architecture models.

Figure 3 shows how a virtual processor in the synchronization layer, like the ones depicted in Fig. 2, is refined. The virtual processor component now

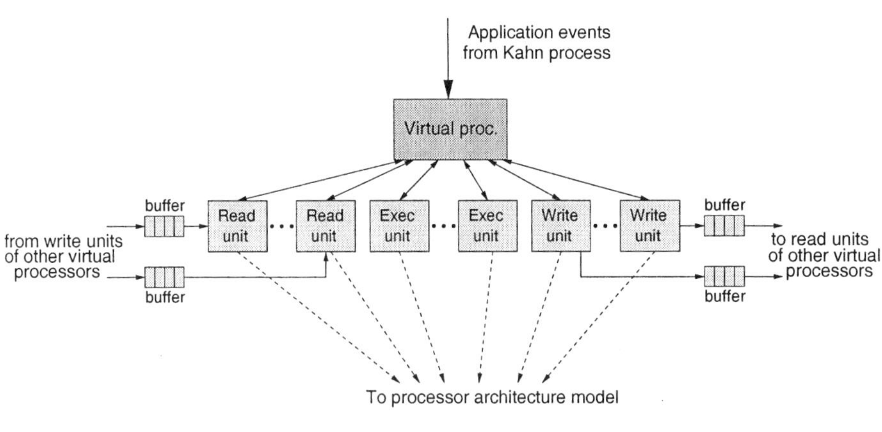

Figure 3 Refining a virtual processor in the synchronization layer.

acts as the front-end to a range of (virtual) functional units. These functional units consist of read, write, and execution units that can operate in parallel. The new virtual processor component has a symbolic-instruction window of parameterizable size in which it stores incoming application events and with which it analyzes them for parallel execution. According to the event type (execute event type, channel from/to which is read/written, etc.), the virtual processor dispatches incoming events to the appropriate functional unit. The number of entries in the symbolic-instruction window limits the number of outstanding (dispatched but not finished) events in the virtual processor. A window size of one implies sequential handling of the application events. In Fig. 3, the arrows from the functional units back to the virtual processor refer to the acknowledgments the functional units transmit whenever the simulation of an event has finished.

The read and write units are connected via buffers* with other virtual processors (as discussed in Section IV), in order to establish the modeling of synchronizations between Kahn application processes in accordance to their mapping onto the underlying architecture. Thus, the read and write units do not dispatch a communication event to the architecture model unless it is safe to do so (i.e., the event cannot cause a deadlock). In addition, the execution and write units do not dispatch their incoming application events to the architecture model before all dependencies for these events are resolved. We will elaborate on this issue in the next section, which discusses the internal synchronizations within a refined virtual processor component.

Figure 4 illustrates how the refined virtual processors can be mapped onto a processor component in the architecture model, which has been refined as well. The read units from the virtual processors that are mapped onto the same processor at the architecture level, are connected to the read units of the processor in the architecture model. Likewise, the virtual execution units are connected to the execution units of the processor architecture model, and so on. The functional units in the architecture model may again be black-box models which sequentially account for the timing consequences of the incoming application events dispatched by the synchronization layer. Alternatively, they may also be further refined. For example, a refined execution unit may model internally pipelined execution of execute events. Furthermore, in the example of Fig. 4 all communication units in the architecture model are connected to a bus model. In reality, communication units within the architecture model may have different connections with each other (directly across a bus or via shared memory, point-to-point, etc.).

---

* Per read or write unit, there may be multiple buffers connected.

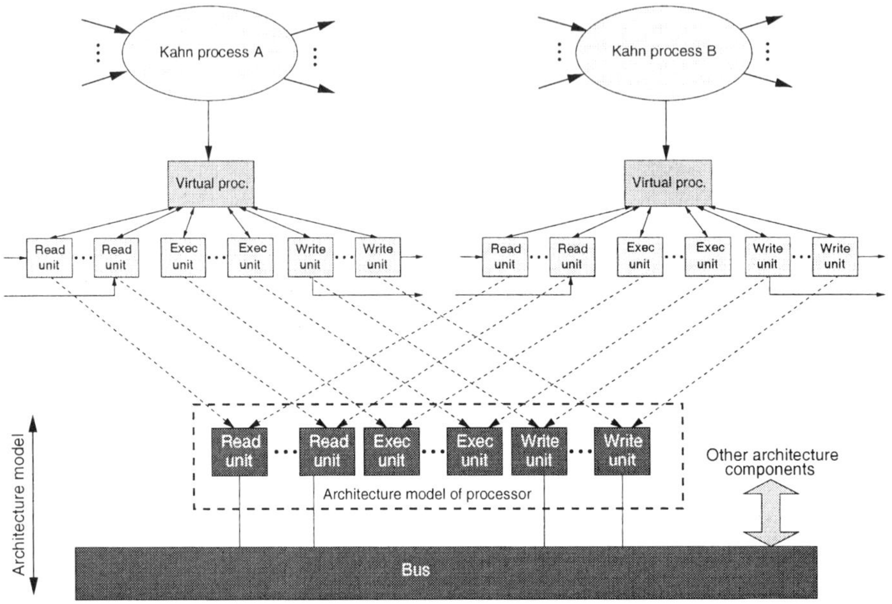

Figure 4 Mapping multiple-refined virtual processors onto a refined processor architecture model.

VI. DATAFLOW FOR FUNCTIONAL UNIT SYNCHRONIZATION

To properly model parallel execution of application events from a single event trace, the dependencies between the events should be taken into account. For example, an execution unit in the synchronization layer may only dispatch an execute event to the execution unit in the architecture model when the read events it depends on have been simulated and delivered the required input for the execute. Likewise, a write event may be dispatched to the architecture model when it is safe to do so and when the read/execute events it depends on have been simulated.

Consider the example in Fig. 5a in which a virtual processor is shown for a processor architecture with a pipeline of two read units, one execution unit, and two write units. In this example, the trace generating Kahn process reads/writes from/to two channels, which are mapped onto separate read and write units. The execute events in this example are dependent on the two preceding read events, while the two write events are dependent on the preceding execute event. In Fig. 5b the resulting (intra-task) pipeline parallelism is illustrated where communication is overlapped with computation.

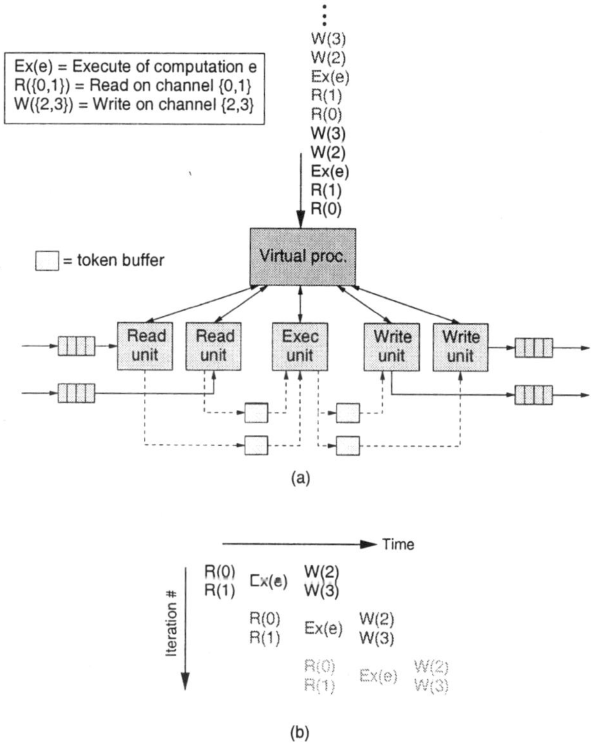

Figure 5 Dataflow-based synchronization to resolve dependencies between functional units in a virtual processor. The architecture shown in (a) exploits pipeline parallelism, which is illustrated in (b).

The synchronization between the functional units in order to resolve dependencies is done via buffered token channels. In Fig. 5a, for example, the read units have a token channel to the execution unit. A read unit sends a token along its token channel whenever a read event is finished (i.e., has been simulated at architecture level). The size of the token channel's buffer determines how far the read unit can run ahead, or in other words, the amount of internal buffering a read unit has. If the token channel's buffer is full, then the read unit stalls until the execution unit has removed one or more tokens from the channel's buffer. During such a stall, a read unit cannot handle new read events.

In our example, the execution unit reads the tokens generated by the read units. Associated with each execute event type, there are two *bitmaps*. The first describes on which token channels the particular execute event is dependent (i.e., which read units produce data needed by the execute event).

The second bitmap describes which functional units are dependent on the execute event. So, it relates to output token channels.

The execution unit must have received a token from all of the required token channels, implying that dependencies have been resolved, before the execute event may be dispatched to the architecture model. Likewise, after an execute event has been simulated at the architecture level, the execution unit sends tokens along the required output token channels (as specified by the second bitmap). As a consequence, the write units, which are waiting for tokens from the execution unit, are enabled to dispatch dependent write events to the architecture model. To summarize, synchronizations due to dependencies between functional units in the synchronization layer are handled using the dataflow principle with token transmissions between the functional units. To be more specific, this dataflow mechanism adheres to integer-controlled dataflow [7]. Of course, the placement of token channels between functional units and their buffer sizes are freely adjustable. For the time being, however, we slightly restricted the choice of functional units as we currently assume that there can be only one execution unit per processor. In Section VIII, we come back to this issue and indicate how our modeling concepts may be extended to support multiple execution units per processor.

To give an impression of what the implemented models look like, Fig. 6 shows the Pearl code for a read unit from the synchronization layer (the variable declarations have been omitted). As Pearl is an object-based language and architecture components are modeled by objects, the code shown in Fig. 6 embodies the class of read unit objects.

```
class v_read_unit

[...]

sig_room : ()
{ }

read : ()
{
   block(sig_room);                 // block until there's room in token buffer
   input_buffer ! get();            // model fetching of data from input FIFO
   ex_unit !! sig_data(unit_id);    // send token to execution unit
   virt_proc !! op_done();          // signal completion to virt. processor
}

{
   while (1) {
      block( read, sig_room );      // main loop
   }
}
```

Figure 6 Pearl code for a read unit object from the synchronization layer.

In its main loop, the read unit object waits for (using the `block ()` primitive) either one of two methods called: `sig_room` or `read`. The `sig_room` method is called whenever there is room for a new token in the token buffer that is associated with the read unit. Multiple calls to this method are queued by the Pearl runtime system. The `read` method is called when a read event needs to be processed by the unit. This method first checks if there is room in the token buffer by waiting until there is at least one call to the `sig_room` method queued up. It then synchronously (!) calls the `get` method in the input buffer object that is connected to the read unit. This means that the read unit will block in virtual time until it receives an acknowledgment from the input buffer object, signaling the end of the data retrieval. Hereafter, the execution unit is signaled by means of an asynchronous method call (!!) to inform it on the availability of the data (i.e., a token is sent). Finally, the virtual processor is signaled that the read unit is ready to receive a new read request. A more thorough explanation of the code is beyond the scope of this chapter. Therefore, the interested reader is referred to [19] for a more detailed discussion of a Pearl code sample.

In our implementation, it is straightforward to change the policy defining when token buffers can be read from or written to. More specifically, a functional unit can wait until all of its required tokens are available before it retrieves the tokens from the buffers or it can retrieve a required token whenever it becomes available. In the latter case, the producer of the token may be unblocked earlier and thereby allowing it to proceed with processing new application events.

We note that the synchronizations between functional units are only performed in the synchronization layer and are not needed within the underlying architecture model. This is because once application events are dispatched from the synchronization layer to the architecture model, they are safe to simulate (i.e., they cannot cause deadlocks and their dependencies have been resolved). This scheme nicely fits our approach, in which all synchronization overheads are accounted for in the synchronization layer.

VII. A CASE STUDY: QR DECOMPOSITION

To validate the previously presented concepts on how to model the exploitation of intra-task parallelism, we performed a case study using a set of application model instances of the well-understood QR decomposition algorithm. These application models were the result of the Compaan work [15] done at Leiden University. The Compaan tool is able to automatically generate Kahn application models from nested-loop programs written in Matlab, which in our case is the QR decomposition algorithm. In addition, it can perform code transformations such as loop unrolling to increase task-level parallelism inside applications [22].

The Kahn application models generated by the Compaan tool are suitable for a direct implementation in hardware on an FPGA. For this purpose, application models are translated into VHDL [11]. This gives us the unique opportunity to validate our abstract architecture models against an actual FPGA implementation. In the VHDL implementation of a Kahn application model, pre-defined node components are connected in a network. This is done according to the connections between the processes in the application model. The node components, which represent the functional behavior of the Kahn processes in the application model, are implemented in a pipelined fashion that is similar to the one shown in Fig. 5. Conceptually, this means that each node component contains a number of read and write units and a single execution unit. So, besides exploiting task-level parallelism by the VHDL network of node components, each node component also exploits intra-task parallelism using its internally pipelined architecture.

Regarding the QR application, we studied five different instances of its application model generated by Compaan. In each instance, the loops in the code were unrolled a different number of times. This loop unrolling creates new Kahn processes, thereby increasing the task-level parallelism available in the application [22]. In Figure 7a, the Matlab code for the QR decomposition algorithm—which is based on the iterative Givens Rotations method—is shown. Figure 7b depicts the Kahn application model Compaan generates for this Matlab code when loop unrolling is turned off. Note that the Kahn model does contain processes for input and output routines (e.g., X_in), which were omitted in Figure 7a. Additional information on the Kahn application model of the QR decomposition algorithm can be found in [11]. For each of the application model instances, we described the structure of the application model in YML to be able to run the model with Sesame's application-model execution engine. As a sidenote, it is worth mentioning that the generation of these YML descriptions of the application model instances is performed fully automatically by means of a visitor tool.

Our Sesame architecture model, onto which the QR application model instances are mapped, is similar to the VHDL implementation of a Kahn application model in the sense that it also consists of processor components connected in a network with a topology identical to that of the application model. Each processor component is modeled with our refined (virtual) processor model (see Section V) and uses the pipelined architecture as shown in Figure 5a. Between processor components in the architecture model there are point-to-point FIFO channels.

Recall that the structure of Sesame's architecture models is described in YML. Because of YML's built-in scripting support, this allowed us to construct a generic reusable template for the refined (virtual) processor

```
for k = 1:1:K,
  for j = 1:1:N,
    [r(j,j),t] = vectorize(r(j,j), x(k,j));
    for i = j+1:1:N,
      [r(j,i),x(k,i),t] = rotate(r(j,i), x(k,i), t);
    end
  end
end
```

(a)

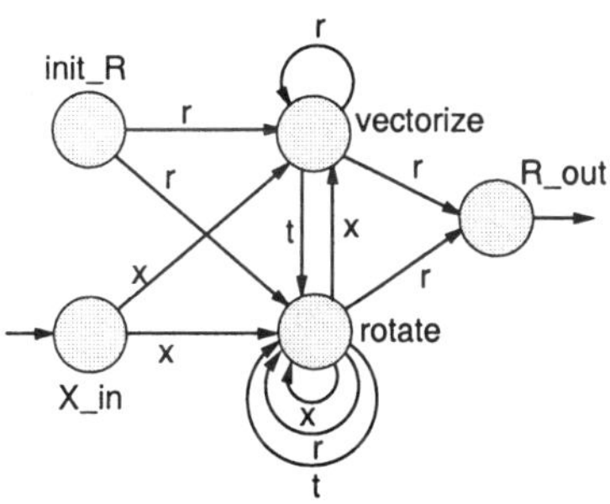

(b)

Figure 7 In (a), the Matlab code for the QR decomposition is shown, while (b) depicts the Compaan–generated Kahn process network without loop unrolling.

model. The processor network in the architecture model was thus obtained by repetitively instantiating this template with possibly different parameters and linking these processor instances together according to the topology of the application model. This topology information was derived from our YML description of the Kahn application model.

A. Experiments

Our first experiments were performed using a Sesame synchronization layer and architecture model with the following characteristics. The size of the FIFO buffers is 256 elements, which guarantees deadlock-free execution of the studied application model instances [11]. The functional units of processor components as well as the FIFO buffers are modeled as black boxes. Read and write operations to the FIFO buffers take three cycles each as specified in [11], while all execute events* are handled in a single cycle. The latter reflects the performance of a fully utilized internal execution pipeline with a single-cycle

* In the QR application model, the execute events consist of vectorize and rotate operations.

throughput. Moreover, the token channels between the functional units at the synchronization layer have single-entry buffers. This means that the read and execution units cannot produce more than one result before consumption (i.e., they have only limited internal buffering).

In Figure 8a, the performance of the FPGA implementation (modeled in VHDL) of the five QR application instances—with loop unroll factors of one to five—is shown. The figure also shows the performance estimates of our black-box Sesame model for these application model instances. These results

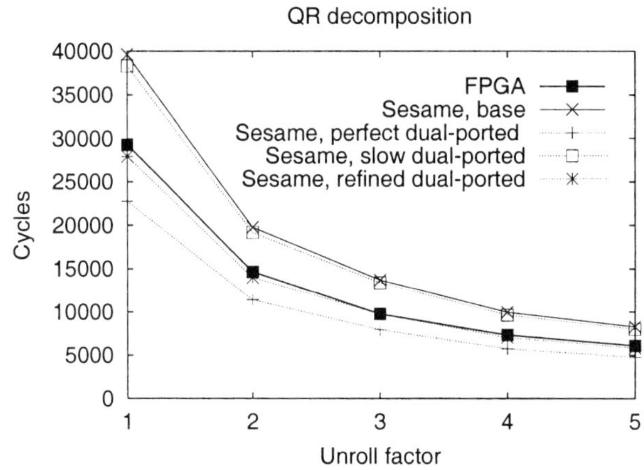

(a)

Difference %	Base model	Dual-ported model		
		Perfect FIFO	Slow FIFO	Refined FIFO
Average	36	-21	32	-3.5
Worst case	40	-22	37	-4.7

(b)

Figure 8 Validation results of our Sesame models for the QR decomposition application against the results from an actual FPGA implementation. The graph in (a) shows the (estimated) performance for five application instances with different loop unroll factors. The table in (b) shows the differences (in %) between estimates from our models and the FPGA numbers.

are referred to as the *base* model in Fig. 8. As shown in Fig. 8b, the black-box model yields an average error of 36% and a worst-case error of 40% with respect to the performance results of the FPGA implementation. The Sesame (base) performance estimates show the correct trend behavior but are consistently more pessimistic than those for the FPGA.

According to [11], the FPGA buffer implementation is based around a dual-ported RAM, where our base model uses single-ported buffers. This explains why the results of the base model are pessimistic. As a next step, we "opened up" the black-box FIFO model and adapted it to include dual-ported behavior. To this end, we modeled three variants of dual-ported FIFO buffers. Two of these variants represent implementation extremes, while the third one reflected the performance behavior of the actual FPGA implementation. The results of these three dual-ported FIFO models are also shown in Fig. 8. The curve labeled *perfect dual-ported* shows the performance estimates when modeling the FIFO buffers as being perfectly dual-ported. The latter means that read and write operations on a buffer can be performed entirely in parallel, even when the buffer is empty. So, when receiving a read request in the empty buffer state, the read is blocked until a write request comes in, after which the incoming (written) data is immediately forwarded to the reading party. Consequently, both read and write latencies are entirely overlapped.

At the other extreme, the curve labeled *slow dual-ported* in Fig. 8 shows the Sesame performance estimates when modeling dual-ported FIFO buffers that are entirely sequential at the empty state. So, when receiving a read request in the empty buffer state, the read is blocked until a write has occurred and finished writing its data into the buffer (in our model, this takes three cycles).

Finally, the curve labeled *refined dual-ported*, shows the Sesame results when incorporating more detailed knowledge of the actual FPGA buffer implementation into our model. Details of the FPGA implementation indicated that a monolithic three-cycle read/write latency for the FIFO buffers does not reflect the actual behavior. In reality, the throughput at both sides of a FIFO buffer is one operation per three cycles, while the read latency turned out to be only one cycle. In our *refined dual-ported* model we have therefore split the three cycle delay into three one-cycle delays and placed them at the appropriate places according to specification of the FPGA buffer implementation. This means that we refined the timing within our model while keeping its abstract structure intact.

Three important conclusions can be drawn from the results in Fig. 8. First, the results reconfirm the modeling flexibility of Sesame. This is because we were able to model the three dual-ported buffer designs by changing less than ten lines in the code of the base model. Second, the results from the "perfect" and "slow" models—representing the two FIFO buffer implementation extremes—immediately indicate that the average accuracy of Sesame's performance estimates must lie in the range of −21% and +32%. In fact, our "refined" model demonstrates how close our performance estimates can approximate reality since it yields an average error of only 3.5% and a worst case error of 4.7%. Knowing that Sesame targets performance evaluation in an early design stage and therefore models at a high level of abstraction, these accuracy numbers are very promising. Third, our results indicate that the studied hardware implementations of the QR decomposition application are highly sensitive to different FIFO buffer designs. Since the performance estimates of the "perfect" buffer model show a speedup of 68% over the results of the "slow" buffer model, the handling of the empty state in the FIFO buffer seems to be an important design issue.

Since Sesame targets performance evaluation in an early design stage (where the design space that needs to be explored typically is very large) the required modeling effort and the simulation speed of Sesame is worth noting. The architecture models in this case study, including the components in the synchronization layer, consist of less than 400 lines of Pearl code. It takes Sesame about 16 seconds on a 333MHz Sun Ultra 10 to perform the architecture simulation for all five application model instances in one batch.

VIII. DISCUSSION

So far, we have assumed that in the set of functional units of a refined (virtual) processor there is only one execution unit. Processing cores, however, may have multiple execution units that can perform computations in parallel. We are currently investigating whether or not our dataflow approach is sufficient for dealing with dependencies between execution units. In any case, for such inter-execution dependencies we need to extend our dataflow scheme such that tokens are typed, as in the tagged-token model [3]. With typed tokens, an execution unit can differentiate between the production of results from different execute event types. To support such typed tokens, the bitmaps need to be extended from single-bit values to multiple-bit values to be able to specify which token types are required for an application event.

Moreover, we currently use static bitmaps per execute event type. We found, however, that this causes problems when, for example, execute events of the same type require data from different read units in different stages of the application model's execution. This can be solved by dynamically adding the bitmap information to the execute events in the traces.

We also intend to investigate whether (aspects from) the work of [23] can be integrated into Sesame since their mapping approach facilitates easier exposure and specification of intra-task parallelism. This could make the use of explicit bitmaps for execute events entirely redundant.

IX. CONCLUSIONS

In this chapter, we presented the techniques applied by the Sesame modeling and simulation environment to model intra-task parallelism exploited at the architecture level for task-parallel applications. To this end, our processor models were refined to the level of functional units that can operate in parallel and are synchronized to resolve dependencies by means of a dataflow mechanism. Using a case study, in which we were able to compare our simulation results with the results from an actual FPGA implementation, we demonstrated that our modeling methodology is flexible and shows good accuracy.

ACKNOWLEDGMENTS

This research was supported by PROGRESS, the embedded systems research program of the Dutch organization for Scientific Research NWO, the Dutch Ministry of Economic Affairs and the Technology Foundation STW. We thank the Embedded Real-time Computing group at Leiden University, and in particular Ed Deprettere, Todor Stefanov, Bart Kienhuis, Vladimir Živković, and Laurentiu Nicolae, for providing the application models for the QR case study and for their invaluable feedback. In addition, we would like to thank Tim Hariss from QinetiQ Ltd (UK) for providing us with details of the FPGA implementation of the QR application.

REFERENCES

1. Cadence Design Systems, Inc., http://www.cadence.com
2. Innoveda Inc., http://www.innoveda.com
3. Arvind, Gostelow, K. P. (1982). The U-Interpreter. *IEEE Computer*. 15(2):42–49, Feb.
4. Balarin, F., Sentovich, E., Chiodo, M., Giusto, P., Hsieh, H., Tabbara, B., Jurecska, A., Lavagno, L., Passerone, C., Suzuki, K., Sangiovanni-Vincentelli, A. (1997). *Hardware-Software Co-design of Embedded Systems-The POLIS approach*. Kluwer Academic Publishers.
5. Brunel, J. -Y., de Kock, E. A., Kruijtzer, W. M., Kenter, H. J. H. N., Smits, W. J. M. (1999). Communication refinement in video systems on chip. In: Proc. 7th Int. Workshop on Hardware/Software Codesign, pp. 142–146, May.

6. Buck, J., Ha, S., Lee, E. A., Messerschmitt, D. G. (1994). Ptolemy: A framework for simulating and prototyping heterogeneous systems. *Int. Journal of Computer Simulation* 4:155–182, Apr.
7. Buck, J. T. (1994). Static scheduling and code generation from dynamic dataflow graphs with integer valued control streams. In: Proc. of the 28th Asilomar conference on Signals, Systems, and Computers, Oct.
8. Coffland, J. E., Pimentel, A. D. (2003). A software framework for efficient system-level performance evaluation of embedded systems. In: Proc. of the ACM SAC, Embedded Systems track, March.
9. Dreike, P., McCoy J., (1997). Co-simulating software and hardware in embedded systems. *Embedded Systems Programming* 10(6), June.
10. Gupta, R. K., Coelho, C. N. Jr., De Micheli, G. (1992). Synthesis and simulation of digital systems containing interacting hardware and software components. In: Proc. of the Design Automation Conference, pp. 225–230, June.
11. Harriss, T., Walke, R., Kienhuis, B., Deprettere, E. F. (2001). Compilation from Matlab to process network realized in FPGA. In: Proc. of the 35th Asiloar conference on Signals, Systems, and Computers, Nov.
12. Hines, K., Borriello, G. (June 1997). Dynamic communication models in embedded system co-simulation. In: Proc. of the Design Automation Conference, pp. 395–400, June.
13. Kahn, G. (1974). The semantics of a simple language for parallel programming. In: Proc. of the IFIP Congress, p. 74.
14. Kienhuis, B., Deprettere, E. F., Vissers, K. A., van der Wolf, P. (1997). An approach for quantitative analysis of application-specific dataflow architectures. In: Proc. of the Int. Conf. on Application-specific Systems, Architectures and Processors, July.
15. Kienhuis, B., Rijpkema, E., Deprettere, E. F. (2000). Compaan: Deriving process networks from Matlab for embedded signal processing architectures. In: Proc. of the 8th International Workshop on Hardware/Software Codesign (CODES'2000), May.
16. Lahiri, K., Raghunathan, A., Dey, S. (June 2001). System-level performance analysis for designing on-chip communication architectures. *IEEE Trans. on Computer-Aided Design of Integrated Circuits and Systems* 20(6):768–783.
17. Lee, E. A., Neuendorffer, S. (March 2000). MoML-a Modeling Markup Language in XML, version 0.4 Technical Report UCB/ERL M00/8. Berkeley: Electronics Research Lab, University of California.
18. Lieverse, P., van der Wolf, P., Deprettere, E. F., Vissers, K. A. (November 2001). A methodology for architecture exploration of heterogeneous signal processing systems. *Journal of VLSI Signal Processing for Signal, Image and Video Technology* 29(3):197–207. Special issue on SiPS'99.
19. Pimentel, A. D., Polstra, S., Terpstra, F., van Halderen, A. W., Coffland, J. E., Hertzberger, L. O. (2002). Towards efficient design space exploration of heterogeneous embedded media systems. In: Embedded Processor Design Challenges: Systems, Architectures, Modeling, and Simulation. Springer: LNCS 2268. pp. 57–73.

20. Pimentel, A. D., Lieverse, P., van der Wolf, P., Hertzberger, L. O., Deprettere, E. F. (Nov. 2001). Exploring embedded-systems architectures with Artemis. *IEEE Computer* 34(11):57–63.
21. Sriram, S., Bhattacharyya, S. S. (2000). *Embedded Multiprocessors: Scheduling and Synchronization*. Marcel Dekker Inc.
22. Stefanov, T., Kienhuis, B., Deprettere, E. F. (2002). Algorithmic transformation techniques for efficient exploration of alternative application instances. In: Proc. of the 10th Int. Symposium on Hardware/Software Codesign (CODES'02), pp. 7–12, May.
23. Živković, V., van der Wolf, P., Deprettere, E. F., de Kock, E. A. (2002). Design space exploration of streaming multiprocessor architectures. In: Proc. of the IEEE Workshop on Signal Processing Systems (SIPS'02), Oct.

6

Energy Estimation and Optimization for Piecewise Regular Processor Arrays

Frank Hannig and Jürgen Teich
University of Paderborn, Paderborn, Germany

I. INTRODUCTION

Today, low power has become an important design criterion due to all the mobile phones and portable computers. These devices must handle increasingly computational-intensive algorithms like video processing (MPEG4) or other digital signal processing tasks (3G), but they are limited in their power budget. The next generation of ULSI chips will allow implementation arrays of hundreds of 32-bit microprocessors and more on a single die. Thus, parallelization techniques and compilers will be of utmost importance in order to map computational-intensive algorithms efficiently to these processor arrays.

In this context, this chapter deals with the specific problem of estimating the power consumption when mapping a certain class of loop-like computations called *piecewise regular algorithms* [28] onto a dedicated processor array. This work may be classified to the area of loop parallelization in the polytope model [10,19].

The rest of the chapter is structured as follows. In Section II, a brief survey of previous work on low power is presented. Section III introduces the class of algorithms we are dealing with. In Section IV, we examine the power consumption of functional units in dependence on their input activity. Afterwards, an energy estimation methodology (when mapping regular algorithms to processor arrays) is described. The methodology and some results are discussed in Section V. In Section VI, a methodology to find energy optimal space-time mappings is proposed. Future extensions and concluding remarks are presented in Section VII.

II. RELATED WORK

A lot of previous work in the area of low power design during high-level synthesis has dealt with the issue of power estimation. Various methodologies for generating accurate models for datapath power consumption were presented.

In general these power estimation techniques can be divided into simulative and non-simulative categories. The non-simulative method in [20] estimated the power consumption from an information theoretical point of view. In [18], the authors described a strategy called a dual bit type (DBT) model where not only the random activity of the least significant bits but also the correlated activity of the most significant bits is taken into account. The method in [12] proposed a modeling approach for functional units that are typically used in digital signal processing systems, such as adders, multipliers, and delay elements. Thereby, a 4-dimensional table-based [11] macro model was used by the authors.

The work of Chandrakasan et al. [3] focuses on transformations at the algorithmic and the architectural level to obtain low power designs. In [2], transformations for nested-loop programs are discussed. In [4,22,23,25], several scheduling and binding techniques for low power are studied. Some energy estimation techniques for processor arrays with hierarchical memory structures are outlined in [8].

In [7] the authors present an approach for energy/power estimation of partioned processor arrays. They focused on finding energy optimal tile sizes and clusterings.

However, to the best of our knowledge, our work presented here is the first that considers the relationship between space-time mappings of computation intensive algorithms and power/energy consumption.

This work is a continuation of the works in [14] and [15]. Here, we specify a power–consumption model used in the methodology described afterwards for energy estimation and optimization of piecewise regular processor arrays.

III. NOTATION AND BACKGROUND

A. Algorithms

The class of algorithms we are dealing with in this chapter is a class of recurrence equations defined as follows.

Definition III.1 (Piecewise regular algorithm). A piecewise regular algorithm contains N quantified equations

$$S_1[I], \ldots, S_i[I], \ldots, S_N[I]$$

Each equation $S_i[I]$ is of the form

$$x_i[I] = f_i(\ldots, x_j[I - d_{ji}], \ldots)$$

where $I \in \mathcal{I}_i \subseteq \mathbb{Z}^n$, $x_i[I]$ are indexed variables, f_i are arbitrary functions, $d_{ji} \in \mathbb{Z}^n$ are constant data dependence vectors and "..." denote similar arguments. The domains are called index spaces and, in our case, are defined as follows.

Definition III.2 (Linearly bounded lattice). A linearly bounded lattice denotes an index space of the form.

$$\mathcal{I} = \{I \in \mathbb{Z}^n \mid I = M\kappa + c \wedge A\kappa \geq b\}$$

where $\kappa \in \mathbb{Z}^l$, $M \in \mathbb{Z}^{n \times l}$, $c \in \mathbb{Z}^n$, $A \in \mathbb{Z}^{m \times l}$ and $b \in \mathbb{Z}^m$. $\{\kappa \in \mathbb{Z}^l \mid A\kappa \geq b\}$ defines an integral convex polyhedron or in case of boundedness a polytope in $\mathbb{Z}^l$. This set is affinely mapped onto iteration vectors I using an affine transformation ($I = M\kappa + c$).

Throughout this chapter, we assume that the matrix M is square and invertible. Then, each vector κ is uniquely mapped to an index point I. Furthermore, we require that the index space is bounded.

For illustration purposes throughout this chapter, the following example is used.

Example III.1. The well-known matrix multiplication algorithm computes the product $C = A \cdot B$ of two matrices $A \in \mathbb{R}^{N_1 \times N_3}$ and $B \in \mathbb{R}^{N_3 \times N_2}$ and is defined as follows.

$$c_{ij} = \sum_{k=1}^{N_3} a_{ik} b_{kj} \quad \forall 1 \leq i \leq N_1 \wedge 1 \leq j \leq N_2.$$

A corresponding piecewise regular algorithm is given by:

Input operations

$$\begin{aligned} &a[i, 0, k] \leftarrow a_{ik} && 1 \leq i \leq N_1 \wedge 1 \leq k \leq N_3 \\ &b[0, j, k] \leftarrow b_{kj} && 1 \leq j \leq N_2 \wedge 1 \leq k \leq N_3 \\ &c[i, j, 0] \leftarrow 0 && 1 \leq i \leq N_1 \wedge 1 \leq j \leq N_2 \end{aligned}$$

Computations

$$\begin{aligned} &a[i, j, k] \leftarrow a[i, j-1, k] && \forall (i\, j\, k)^{\mathrm{T}} = I \in \mathcal{I} \\ &b[i, j, k] \leftarrow b[i-1, j, k] && \forall (i\, j\, k)^{\mathrm{T}} = I \in \mathcal{I} \\ &z[i, j, k] \leftarrow a[i, j, k] \cdot b[i, j, k] && \forall (i\, j\, k)^{\mathrm{T}} = I \in \mathcal{I} \\ &c[i, j, k] \leftarrow c[i, j, k-1] + z[i, j, k] && \forall (i\, j\, k)^{\mathrm{T}} = I \in \mathcal{I} \end{aligned}$$

Output operations

$$c_{ij} \quad \leftarrow c[i,j,N_3] \qquad 1\leq i \leq N_1 \wedge 1\leq j \leq N_2$$

The data dependence vectors are

$$d_{aa} = (0\ 1\ 0)^{\mathrm{T}}, d_{bb} = (1\ 0\ 0)^{\mathrm{T}}, d_{cc} = (0\ 0\ 1)^{\mathrm{T}},$$
$$d_{az} = (0\ 0\ 0)^{\mathrm{T}}, d_{bz} = (0\ 0\ 0)^{\mathrm{T}}, d_{zc} = (0\ 0\ 0)^{\mathrm{T}}$$

The index space is given by

$$\mathcal{I} = \left\{ I = \begin{pmatrix} i \\ j \\ k \end{pmatrix} \in \mathbb{Z}^3 \mid \begin{pmatrix} 1 & 0 & 0 \\ -1 & 0 & 0 \\ 0 & 1 & 0 \\ 0 & -1 & 0 \\ 0 & 0 & 1 \\ 0 & 0 & -1 \end{pmatrix} I \geq \begin{pmatrix} 1 \\ -N_1 \\ 1 \\ -N_2 \\ 1 \\ -N_3 \end{pmatrix} \right\}$$

Computations of piecewise regular algorithms may be represented by a *dependence graph* (DG). The dependence graph of the algorithm of Example III.1 is shown in Fig. 1a. The dependence graph expresses the partial order between the operations. Each variable of the algorithm is represented at every index point $I \in \mathcal{I}$ by one node. The edges correspond to the data dependencies of the algorithm. They are regular throughout the algorithm (i.e., $a[i,j,k]$ is directly dependent on $a[i,j-1,k]$). The dependence graph specifies implicitly all legal execution orderings of operations: if there is a directed path in the dependence graph from one node $a[J]$ to a node $z[K]$ where $J, K \in \mathcal{I}$, then the computation of $a[J]$ must precede the computation of $z[K]$.

Henceforth, and without loss of generality,* we assume that all indexed variables are embedded in a common index space $\mathcal{I}$. Then, the corresponding dependence graphs can be represented in a reduced form.

Definition III.3 (Reduced dependence graph). A reduced dependence graph (RDG) $G = (V, E, D)$ of dimension n is a network where V is a set of nodes and $E \subseteq V \times V$ is a set of edges. To each edge $e = (v_i, v_j)$ there is associated a dependence vector $d_{ij} \in D \subset \mathbb{Z}^n$.

The RDG of the matrix multiplication algorithm is shown in Fig. 1a. Each node v in the graph corresponds to one equation in the section computations of the algorithm.

* All methods described can also be applied to each quantification individually.

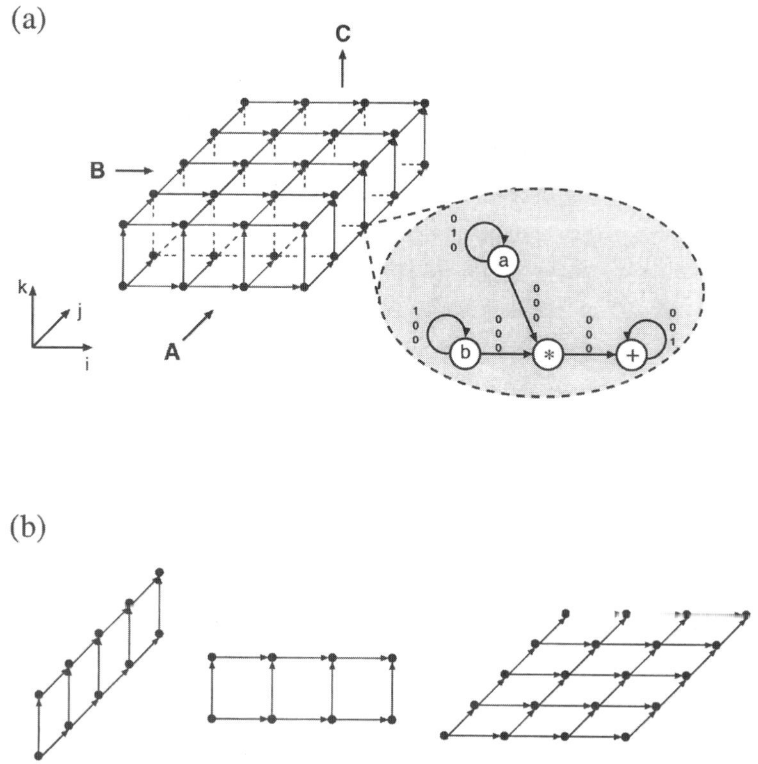

Figure 1 In (a), an index space and the reduced dependence graph is shown. Some possible mappings are depicted in (b).

B. Space-Time Mapping

Linear transformations, as in Eq. (1), are used as *space-time mappings* [16,21] in order to assign a *processor index* $p \in \mathbb{Z}^{n-1}$ (space) and a *sequencing index* $t \in \mathbb{Z}$ (time) to index vectors $I \in \mathcal{I}$.

$$\begin{pmatrix} p \\ t \end{pmatrix} = TI = \begin{pmatrix} Q \\ \lambda \end{pmatrix} I \tag{1}$$

In Eq. (1), $Q \in \mathbb{Z}^{(n-1)\times n}$ and $\lambda \in \mathbb{Z}^{1\times n}$. The main reasons for using linear allocation and scheduling functions is that the data flow between PEs is local and regular, which is essential for low-power VLSI implementations. The interpretation of such a linear transformation is as follows: The set of operations defined at index points $\lambda \cdot I = \text{const}$ are scheduled at the same time step. The index space of allocated processing elements (*processor space*) is denoted by $\mathcal{Q}$ and is given by the set $\mathcal{Q} = \{p \mid p = Q \cdot I \wedge I \in \mathcal{I}\}$. This set can also be obtained by choosing a projection of the dependence graph along a vector $u \in \mathbb{Z}^n$, i.e., any coprime* vector u satisfying $Q \cdot u = 0$ [16] describes the allocation equivalently.

Allocation and scheduling must satisfy that no data dependencies in the DG are violated. This is ensured by the following *causality constraint*, $\lambda \cdot d_{ij} \geq 0 \ \forall (v_i, v_j) \in E$. A sufficient condition for guaranteeing that no two or more index points are assigned to a processing element at the same time step is given by

$$\operatorname{rank}\begin{pmatrix} Q \\ \lambda \end{pmatrix} = n \tag{2}$$

Using the projection vector u satisfying $Q \cdot u = 0$, this condition is equivalent to $\lambda \cdot u \neq 0$ [28].

Definition III.4 (Iteration interval). [30]. *The iteration interval π of an allocated and scheduled piecewise regular algorithm is the number of time instances between the evaluation of two successive instances of a variable within one processing element.*

Definition III.5 (Block pipelining period). [17]. *The block pipelining period of an allocated and scheduled piecewise regular algorithm is the time interval between the initiations of two successive problem instances and is denoted by β.*

Let us consider the matrix multiplication algorithm introduced in Example III.1 as a problem instance. The whole matrices A and B have to be read into the processor array before the next pair can be read, the time between these input operations is the block pipelining period β. Let λ be the schedule vector. Then, the block pipelining period β may be computed as follows.

$$\beta = \max_{I_1 \in \mathcal{I}}\{\lambda \cdot I_1\} - \min_{I_2 \in \mathcal{I}}\{\lambda \cdot I_2\} = \max_{I_1, I_2 \in \mathcal{I}}\{\lambda(I_1 - I_2)\}$$

IV. POWER MODELING AND ENERGY ESTIMATION

In digital CMOS circuits, the dominant source of power consumption is switching power [26]. The average power consumed by a CMOS gate can be computed using the following equation,

---

* A vector x is said to be *coprime* if the absolute value of the greatest value of the greatest common divisor of its elements is one.

$$P_{sw} = \frac{1}{2} C_L V_{dd}^2 N f$$

where C_L is the gate output load capacitance, V_{dd} is the supply voltage, f is the clock frequency, and N is the average or expected number of output transitions per clock cycle.

Due to the influence of the switching activity on the power consumption, our main idea is to exploit the fact that power consumption is drastically reduced when some inputs of a functional unit remain unchanged for $n > 1$ clock cycles.

Here, we want to discuss the impact of the space-time mapping on the power and energy consumption respectively of the resulting processor array. Our approach identifies regions with decreased switching activity of functional units' input operands and takes these power savings into account. An estimation methodology is presented in the following. This methodology estimates for a given piecewise regular algorithm and a space-time mapping T the average power consumption of the entire array.

Briefly described, this methodology can be subdivided into two hierarchical estimation steps,

PE-level power estimation, and
array-level power estimation.

A. PE-Level Power Estimation

A diagram of the internal structure of a typical processor element is shown in Fig. 2. It consists of a core where all the functional units are located, a controller, and some delay registers. In Section V, we quantify typical per-

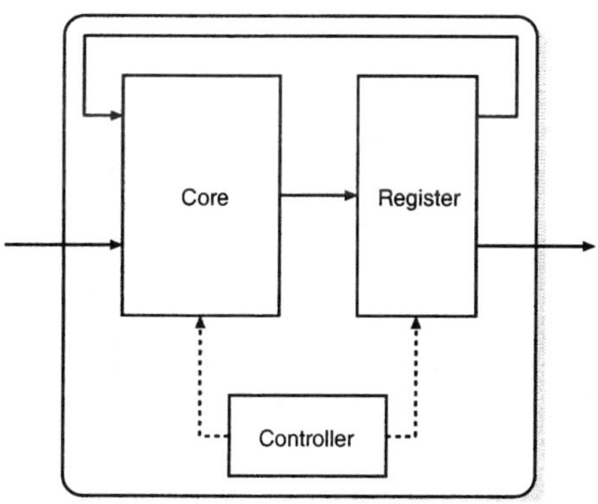

Figure 2 Schematically internal structure of one processor element.

centages of power consumption for the functional units P_{FU}, the control structures P_{Ctrl}, and the registers P_{Rg}, and these parts' proportion of the overall power consumption of one processing element. The power consumption of one PE can be approximated as follows.

$$P_{\text{FU}}(\lambda, u) = P_{\text{FU}}(u) + P_{\text{Ctrl}}(\lambda) + P_{\text{Rg}}(\lambda)$$

For characterization of the functional units (adders, multipliers, etc), standard register-transfer level power estimation tools from Synopsys [27] are used.

In Table 1, the average power consumption of some 16-bit functional units are listed (A = ripple-carry adder, B = carry-save array multiplier, C = carry-save array multiplier with two pipeline stages, and D = Wallace-tree multiplier with three pipeline stages). Each functional unit has two input operands. The value of one operand is assumed to be constant for n clock cycles; the other can change randomly in every clock cycle. These values are shown in Fig. 3a for the 16-bit ripple-carry adder and Fig. 3b for the multipliers, respectively. The curves are derived by regression, where the function is of type $P = a_0 + a_1 e^{-n} + a_2 n e^{-n} + a_3 n^2 e^{-n}$. The regression is good enough to have less than 2% error. Since we are only interested in integer multiples of the clock cycle for n, the derived models may be stored in a table without too much effort.

It can be seen from these figures that the power consumption of a functional unit depends heavily on the number of cycles of one input operand stays constant. A good estimation methodology, therefore, should exploit this observation for obtaining accurate estimations and from studying the influence of space-time mappings on the resulting power consumption.

Table 1 Average Power Consumption of Different Functional Units

n	$P_{avg,A}$	$P_{avg,B}$	$P_{avg,C}$	$P_{avg,D}$
1	26.97 μW	204.2 μW	212.0 μW	319.6 μW
2	22.33 μW	155.4 μW	164.0 μW	225.0 μW
3	18.82 μW	138.6 μW	145.6 μW	190.1 μW
4	16.99 μW	129.6 μW	137.3 μW	175.1 μW
5	16.31 μW	125.4 μW	133.8 μW	164.3 μW
6	15.68 μW	120.5 μW	128.4 μW	159.4 μW
7	15.48 μW	119.5 μW	125.2 μW	153.3 μW
8	15.29 μW	116.8 μW	124.4 μW	151.6 μW
9	15.09 μW	116.3 μW	123.7 μW	147.8 μW
10	14.89 μW	115.5 μW	122.7 μW	145.8 μW
∞0	8.49 μW	—	—	—

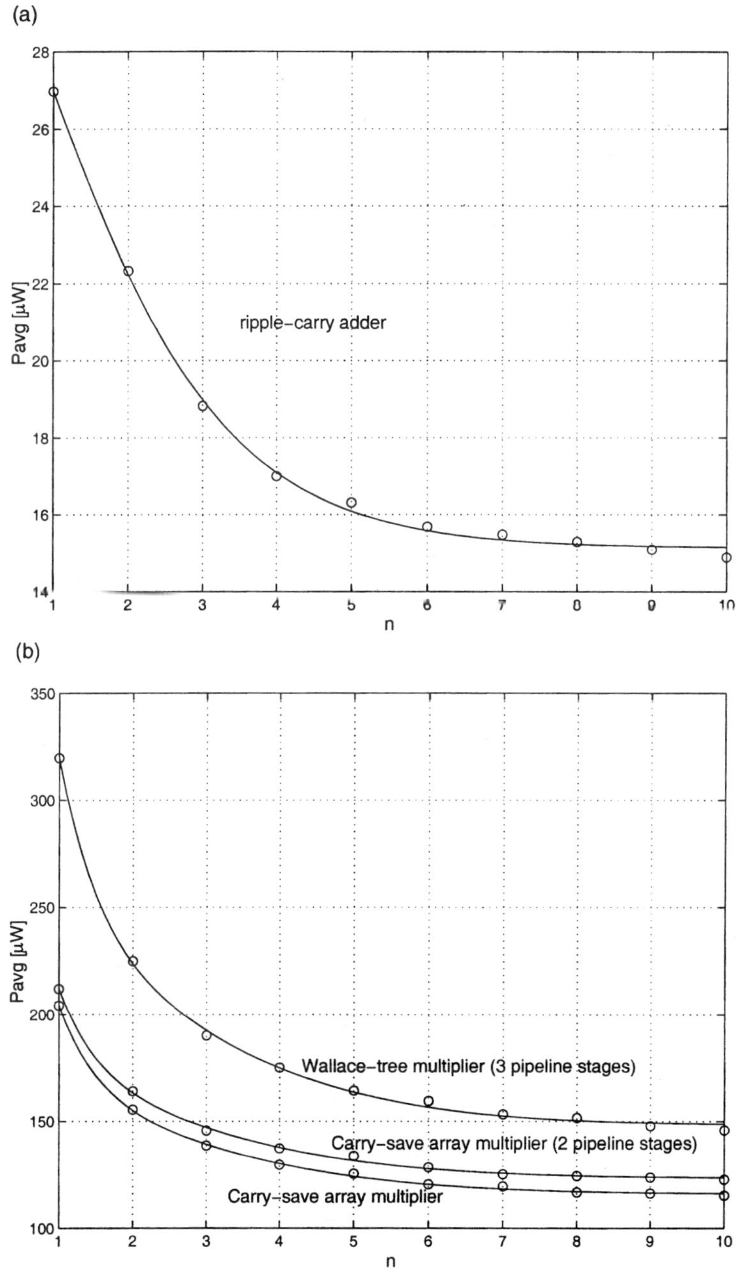

Figure 3 Average power consumption of some 16-bit functional units when one operand is constant for n clock cycles and the other can change randomly in every clock cycle.

B. Array-Level Power Estimation

Based on the class of piecewise regular algorithms, we want to estimate the power consumption for a given space-time mapping $T = (Q\ \lambda)^{\mathrm{T}}$. It is obvious that the cost (number of processor elements) and the latency is influenced by the space-time mapping. In earlier work [13], we described how to determine the cost and the latency as a measure of performance. Here, we briefly outline the main ideas. If we assume that processor arrays are resource-dominant, we are able to approximate the cost as being proportional to the processor count. *Ehrhart polynomials* [6,9] may be evaluated to count the number of points (processor elements, $\#PE$) in the projected index space.

The latency is determined by solving a minimization problem, which may be formulated as a mixed-integer linear program (MILP) [29,30]. Also, modified low power scheduling and binding techniques as in [23,25] can be applied to compute a suited schedule.

Here, we discuss the impact of the space-time mapping on the power and energy consumption, respectively, of the resulting processor array. Our approach identifies regions with decreased switching activity of functional units' input operands and take these power savings into account. An estimation algorithm is presented on the following pages. The algorithm estimates for a given RDG G, an index space $\mathcal{I}$, a space-time mapping T, the number of processor elements $\#PE$, and the block pipelining period β, the average power consumption P_{array} of the entire array. The processor count $\#PE$ and the block pipelining period β of the array may be computed as described earlier in this chapter.

Once the average power consumption P_{array} of the entire processor array is estimated, the energy consumption (per problem instance) is computed as follows,

$$E = \beta \cdot P_{\mathrm{array}}$$

```
POWER ESTIMATION
IN:     RDG G, I, T = (Q λ)^T, #PE, and β
OUT: P_array
BEGIN
  P_PE ← 0
  FOR all nodes v ∈ G DO
    P_v,1 ← lookUpPower(v,1)
    P_PE ← P_PE + P_v,1
  ENDFOR
  P_array ← #PE · P_PE
  FOR all edges e ∈ G DO
    d is dependence vector of edge e
    node v ← source(e)
```

```
13       node w ← target(e)
14       IF (v = w) THEN
15         IF (S_v is propagation equation) THEN
16           IF (Q · d = 0) THEN
17             FOR all adjacent edges e' of v
18               d' is dependence vector of edge e'
19               IF (d' = 0) THEN
20                 w ← target(e')
21                 P_w,1 ← lookUpPower(w, 1)
22                 P_w,β ← lookUpPower(w, β)
23                 P_array ← P_array - #PE · ( P_w,1 - P_w,β)
24               ENDIF
25             ENDFOR
26           ENDIF
27         ELSE
28           (k, m) ← getOperandFixedCycles(T, v)
29           P_w,1 ← lookUpPower(w, 1)
30           P_w,k ← lookUpPower(w, k)
31           P_array ← P_array - m · ( P_w,1 - P_w,k)
32         ENDIF
33       ENDIF
34     ENDFOR
35   END
```

In our experiments, we assumed that the iteration period π is one and that each RDG node is mapped onto a dedicated resource (no resource sharing of functional units). Our estimation algorithm can be subdivided into two phases. In the first phase, the worst case power consumption is computed (i.e., when the switching activity of all functional units' input operands is highest). Therefore, the power consumption P_{PE} of one processor element is determined by summation of the power consumption $P_{v_i,1}$ of all of its FUs

$$P_{\text{PE}} = \sum_{\forall v_i \in V} P_{v_i,1}$$

The one in the term $P_{v_i 1}$ denotes that operands can change in every clock cycle.

Subsequently, the power consumption of the entire array is obtained by extrapolation of this value. In the second phase of the algorithm, array regions with lower switching activity are detected. Therefore, the whole reduced dependence graph is traversed to examine self-loops* (see line 14).

*A self-loop is an edge where source and target node are the same.

These self-loops correspond to inputs of a processor element. If these inputs remain unchanged for more than one period, the switching activity is decreased and consequently also the power. It remains to be determined how many clock cycle inputs are constant and how many processor elements are affected. Two cases can be differentiated.

1. *Propagation equations mapped onto itself.* Propagation equations are used only to distribute data from one processor to another. Due to the regularity and locality of the considered processor arrays, they occur very commonly. If such a propagation equation is mapped onto itself ($Q \cdot d = 0$, see line 16) no data transport is needed (i.e., the data remains in one processor element unchanged for β cycles until the next problem instance is fed into the array). Thus, the switching activity of all adjacent nodes v_i (functional units) in the same processor element is reduced. Therefore, the estimation value of the average power consumption is corrected (decreased) by $P_{v_i 1}$-$P_{v_i,\beta}$. As a propagation equation has global influence, the activity is reduced in every processor element (#*PE*).
2. *Other self-loops.* These are the remaining inputs which may be constant for k clock cycles. Let the number of processor elements with these constant inputs be denoted m. Let $\mathcal{I}_{in_1}$ be the input index space of variable in_i. Transforming this index space by Q and counting the number of points in the transformed space, gives m.

$$m = | \{ I \in \mathbb{Z}^{n-1} \mid I = Q \cdot I_{in_1} \wedge I_{in_1} \in \mathcal{I}_{in_1} \} |$$

This counting problem is similar to the earlier problem described and can be obtained by a geometrical approach [5]. The number of integral points can be determined by consideration if the given projection vector u ($Q \cdot u = 0$) enters a facet of the index space $\mathcal{I}_{in_1}$ and how *thick* this facet must be until two points projected onto each other. Algebraically, the *thickness* is derived from the value of the inner product of the normal vector of a facet and the projection vector. The union of thick facets can be a nonconvex polytope. The number of integral points inside this (non-)convex polytope is determined by the use of *Ehrhart polynomials* [6].

Once k and m are determined (see line 28, function `getOperandFixedCycles`), the overall estimated power consumption value can be improved by subtracting $m \cdot (P_{in_1,1} - P_{in_1,\mathrm{k}})$.

In the next section the overall algorithm is explained and quantitative results are discussed.

V. EXPERIMENTS

Reconsider the introductory Example III.1. As an allocation, we chose a 16-bit ripple-carry adder for the addition and a three-stage pipelined Wallace-tree multiplier for the multiplication. The input operations a and b were mapped each to one resource of type *input*. The execution times of these operations are zero. This is equivalent to a multicast without delay to a set of processors. Furthermore, let $u = (1\ 0\ 0)^T$ be the chosen projection vector. Then, after scheduling and cost calculus, we obtain the schedule vector $\lambda = (1\ 0\ 1)$ and as cost $\#PE = N_2 \cdot N_3$. Now, with this information we are able to estimate the power consumption by applying the proposed algorithm. First, the worst-case power consumption is determined (i.e., the switching activity of functional units' when input operands change each cycle). Second, in the main part of the algorithm, two types of equations with lower input activity are detected and the overall power consumption is refined.

The processor array for a projection in direction $u = (1\ 0\ 0)^T$ is shown in Fig. 4. Due to this projection, the variable b is mapped onto itself. From this it follows tht one operand of the multiplication remains unchanged for some time. At the beginning of a computation, the whole matrix B is input simultaneously to the array, whereas the matrix A is fed sequentially row by row from the left side into the array. Since the matrix A has N_1 rows, one operand of the multiplier is fixed for $\beta = N_1$ clock cycles, which significantly reduces the power consumption in the multipliers by 45% (see Table 1). On account of the design regularity the power savings can multiplied by $\#PE$ (line 23 of the algorithm). The second point where less power is consumed is the constant input variable c. One input of the adders in the lower row of the

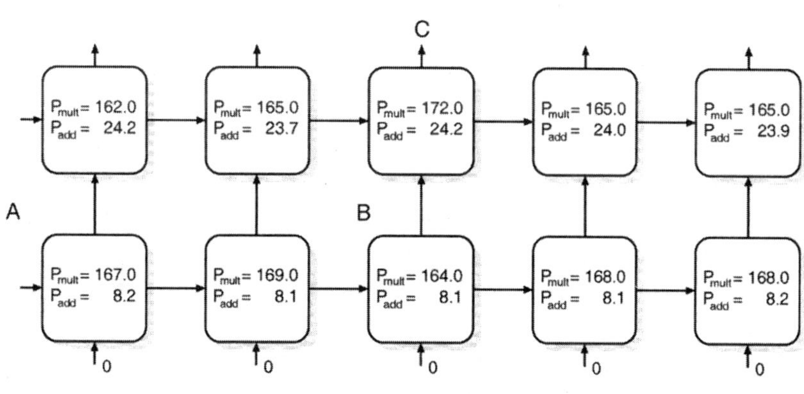

Figure 4 Processor array for $u = (1\ 0\ 0)^T$, $N_1 = 4$, $N_2 = 5$, and $N_3 = 2$.

processor array is permanently zero. These partial regions with reduced power consumption in the array are determined by the function `getOperandFixedCycles`. In addition to the time ($k=\infty$) where one input remains unchanged, the number $m=N_2$ of processors with reduced switching activity is returned.

In Table 2, the power consumption for different projection vectors is shown, where for illustration purposes, the upper boundaries of the index space are set to $N_1 = 4$, $N_2 = 5$, and $N_3 = 2$. In the table, P_{sim} is the exact value obtained by simulation of the entire array. The worst case extrapolation (line 4–9 in the algorithm) is denoted by P_{ext}. The power consumption of our estimation algorithm is labeled with P_{est}. Where the simple extrapolation method has errors up to 81%, our approach is very accurate with errors less than 5%.

Furthermore, the energy values per matrix multiplication in the table show the significant influence of the chosen space-timing mapping. Different mappings can lead to energy consumptions that can differ up to a factor of two.

A. Quantification of the Power Consumption Inside One Processor Element

In this subsection, we quantify the percentages of power consumption for the functional units P_{FU}, the control structures P_{Ctrl}, and the registers P_{Rg}. In Table 3, the proportions of these parts to the overall power consumption of one processing element for the three unit vector mappings are depicted. For the matrix multiplication algorithm the major part of the power consumption is caused by the functional units, this part is around 90%. Where the power consumption of the registers is only 4.1–6.6% of the total power consumption of one processor element. It should be recalled that the iteration interval is only one and no resource sharing is used since at each index point only one multiplication and one addition has to be performed.

Table 2 Average Power and Energy Consumption of Different Mappings

u	P_{sim} [μW]	P_{ext} [μW]	Err_{ext} [%]	P_{est} [μW]	Err_{est} [%]	E_{sim} [pJ]	E_{est} [pJ]
$(1\ 0\ 0)^T$	2020	3466	71.6	1928	−4.6	80.8	77.1
$(0\ 1\ 0)^T$	1530	2773	81.2	1456	−4.8	76.5	72.8
$(0\ 0\ 1)^T$	7260	6931	−4.5	6931	−4.5	145.2	138.6

Table 3 Percentages of Power Consumption for Functional Units, Control Structures, and Registers

Algorithm	u	$P_{FU}[\%]$	$P_{Ctrl}[\%]$	$P_{Rg}[\%]$
Matrix multiplication	$(1\ 0\ 0)^T$	87.8	5.7	6.6
Matrix multiplication	$(0\ 1\ 0)^T$	86.6	7.1	6.3
Matrix multiplication	$(0\ 0\ 1)^T$	91.8	4.1	4.1
LU decomposition (A)	$(1\ 0\ 0)^T$	78.5	9.2	12.3
LU decomposition (B, C)	$(1\ 0\ 0)^T$	88.0	6.1	5.9

The second example is a piecewise regular algorithm for LU decomposition. In Fig. 5, a piecewise regular processor array for the LU decomposition is schematically shown. This array can be subdivided into three pieces, where the Parts A and B also change their functionality over the time.

Since the Part B and C divisions are performed the percentage of functional unit of the overall power consumption is greater than for the Part A. The percentages for the different parts of the LU decomposition array are listed in Table 3.

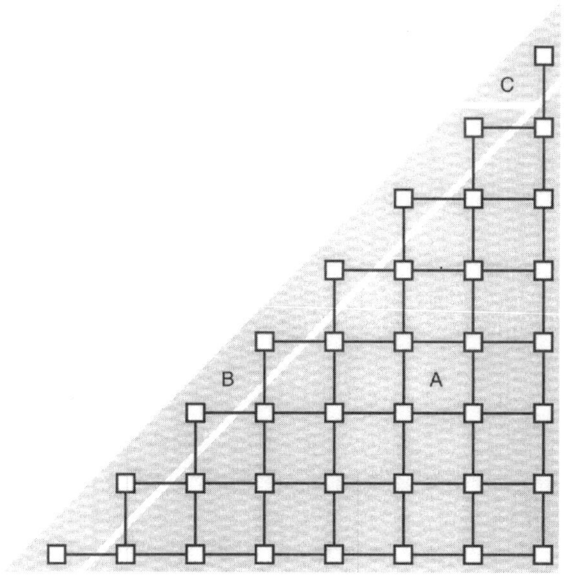

Figure 5 Sketch of piecewise regular processor array for LU decomposition.

VI. DETERMINATION OF ENERGY-OPTIMAL SPACE-TIME MAPPINGS

In the previous section we showed that different space-time mappings have a great influence on energy consumption. In this section we want to make use of this fact to determine energy-optimal space-time mappings.

The algorithm proposed in Section IV.B determines for an arbitrary given space-time mapping $T = \binom{Q}{\lambda}$, $T \in \mathbb{Z}^{n \times n}$ the power consumption. Since T can equivalently described by a schedule vector $\lambda \in \mathbb{Z}^{1 \times n}$ and a projection vector $u \in \mathbb{Z}^n$ we have $2n$ parameters of possible space-time mappings. In [13], we presented efficient pruning techniques for the search of optimal space-time mappings (projection vectors). Here, we summarize the main ideas: (1) only consider co-prime projection vectors u, and (2) only consider co-prime vectors that have the properties that at least two points in $\mathcal{I}$ are projected onto each other. This leads to a search space of co-prime vectors in a convex polytope called difference-body of points in $\mathcal{I}$. Finally, in this reduced search space, we can exploit symmetry to exclude search vectors $v = -v'$ such that typically, only few projection vector candidates v have to be investigated. *Ehrhart polynomials* [6,9] may be evaluated to count the number of points in the projected index space. Let $|U|$ be the number of projection vector candidates and for each projection vector $u \in U$ the minimal latency was determined by solving a mixed integer linear program (MILP) [13,29,30]. Then, we must also estimate for $|U|$ space-time mappings the power consumption. Since the set U can still be very large we propose an efficient heuristic methodology to find an energy-optimal space-time mapping in the following.

```
ENERGY OPTIMIZATION
IN:     RDG G, I
OUT:  E_opt, T_opt
BEGIN
  U ← Ø
  FOR all edges e ∈ G DO
    d is dependence vector of edge e
    IF (d ≠ 0 ∧ d ≠ U) THEN
      use d as projection vector and construct Q from it
      #PE ← determineNoOfPEs(d)
      (λ, β) ← minimizeLatency(G, I, d)
      P_array ← powerEstimation(G, I, T, #PE, β)
      E ← β · P_array
      IF (E < E_opt) THEN
        E_opt ← E
```

```
            T_opt ← T
         ENDIF
         U ← U ∪ d
      ENDIF
   ENDFOR
END
```

To obtain energy-optimal space-time mappings we use the same observation when identifying regions with decreased switching activity of functional units' input operands to get an accurate power estimation, therefore, self-loops are examined. When such a self-loop is projected onto itself the switching activity may be reduced because if it is a matter of (a) propagation equation mapped onto itself, the data propagated remains in one processor element unchanged for β clock cycles or (b) other self-loops, the data remains unchanged for m clock cycles in dependence on the input space size. Due to these observations we need only to examine projections (mappings) which project one (or more) self-loops onto itself, Energy-optimal space-time mappings can be determined by the above algorithm.

Since a given algorithm consists of N quantified equations in the worst case, only N times the energy must be estimated. Once the power tables of the functional units are generated, this can be done within seconds.

VII. CONCLUSIONS AND FUTURE WORK

A first study of a matrix multiplication algorithm has shown great impact in a chosen mapping on the average energy consumption. The resulting array and the accuracy (errors $< 5\%$) of our estimation approach is superior when compared with RTL power estimation tools from Synopsys [27].

Furthermore, our methodology is independent of the problem (array) size, since, an estimation with Synopsys design tools has linear time and memory complexity in dependence on the number of processor elements. Power estimation for large processor arrays using the Synopsys design tools rapidly become crucial since memory usage is growing to GBytes and estimation time to several hours. Exact comparisons of the complexity and also a quantification of the percentages of power consumption of the functional units, the controller, and the registers with respect to the overall power consumption of one processing element were presented also. First experiments of matrix multiplication and LU decomposition have shown that since all data is stored locally inside PE's registers, the part of the register power consumption averages from ~ 4–12% of the overall power consumption. Furthermore, we presented a fast heuristic algorithm to find energy-optimal space-time mappings.

Our new estimation methodology is currently integrated into the PARO design system and can be used during the process of automated synthesis of regular circuits. PARO is a design system project for modeling, transforming, optimization, and processor synthesis for the class of piecewise linear algorithms [1,24].

In the future, we would like to use the estimation technology presented here and extend it also to reduced (i.e., partioned arrays). First results on the computation of power in reduced size arrays have been recently presented in [7]. We want to combine our knowledge about the influence of a chosen mapping on the average energy consumption with this work.

ACKNOWLEDGMENTS

This work was supported in part by the German Science Foundation (DFG) Project SFB 376 "Massively Parallel Computation: Algorithms, Design Methods, Applications."

REFERENCES

1. Bednara, M., Teich, J. (2001). Synthesis of FPGA Implementations from Loop Algorithms. In: First Internationl Conference on Engineering of Reconfigurable Systems and Algorithms (ERSA'01), Las Vegas, NV, pp. 1–7, June.
2. Catthoor, F., Franssen, F., Wuytack, S., Nachtergaele, L., De Man, H. (1994). Global Communication and Memory Optimizing Transformations for Low Power Systems. In: VLSI Signal Processing Workshop,. pp. 178–187, October.
3. Chandrakasan, A. P., Potkonjak, M., Mehra, R., Rabaey, J., Brodersen, R.W. (January 1995). Optimizing Power Using Transformations. *IEEE Transactions on Computer-Aided Design of Integrated Circuits and Systems* 14(1):12–31.
4. Chang, J. -M., Pedram, M. (1996). Module Assignment for Low Power. In: IEEE European Design Automation Conference (EuroDAC). Geneva, Switzerland, pp. 376–381, September.
5. Clauss, P. (1996). Counting Solutions to Linear and Nonlinear Constraints through Ehrhart polynomials: Applications to Analyse and Transform Scientific Programs. In: Tenth ACM International Conference on Supercomputing. Philadelphia, Pennsylvania, May.
6. Clauss, P., Loechner, V. (July 1998). Parametric Analysis of Polyhedral Iteration Spaces. *Journal of VLSI Signal Processing* 19(2):179–194.
7. Derrien, S., Rajopadhye, S. (2002). Energy/Power Estimation of Regular Processor Arrays. In: 15th International Symposium on System Synthesis (ISSS'02). Kyoto, Japan, October.
8. Eckhardt U. (2000). Algorithmus-Architektur-Codesign für den Entwurf digi-

taler Systeme mit eingebettetem Prozessorarray und Speicherhierarche. PhD thesis, Technische Universität Dresden, Fakultät Elektrotechnik, Dresden, Germany.
9. Ehrhart, E. (1977). Polynômes arithmétiques et Méthode des Polyèdres en Combinatoire, volume 35 of International Series of Numerical Mathematics. (1 edition). Basel: Birkhäuser, Verlag.
10. P. Feautrier (1996). Automatic Parallelization in the Polytope Model. Technical Report 8, Laboratoire PRiSM, Université des Versailles St-Quentin en Yvelines, 45, avenue des États-Unis: F-78035. Versailles Cedex, June.
11. Gupta, S., Najm, F. N. (1999). Power Macro-Models for DSP Blocks with Application to High-Level Synthesis. In: IEEE International Symposium on Low Power Electronics and Design. San Diego, CA, pp. 103–105, August.
12. Gupta, S., Najm, F. N. (February 2000). Power Modeling for High-Level Power Estimation. *IEEE Transactions on Very Large Integration (VLSI) Systems* 8(1):18–29.
13. Hannig, F., Teich, J. (September 2001). Design Space Exploration for Massively Parallel Processor Arrays. In: Malyshkin, V., ed. *Parallel Computing Technologies, 6th International Conference, PaCT 2001, Proceedings, volume 2127 of Lecture Notes in Computer Science (LNCS).* Novosibirsk, Russia: Springer, pp. 51–65.
14. Hannig, F., Teich, J. (2002). Energy Estimation for Piecewise Regular Processor Arrays. In: Proceedings of the Second International Samos Workshop on Systems, Architecture, Modelling, and Simulation (SAMOS 2002). Island of Samos, Greece, July.
15. Hannig, F., Teich, J. (2002). Energy Estimation of Nested Loop Programs. In: Proceedings 14th Annual ACM Symposium on Parallel Algorithms and Architectures (SPAA 2002), Winnipeg, Manitoba, Canada. ACM Press, August.
16. Kuhn, R. H. (1980). Transforming Algorithms for Single-Stage and VLSI Architectures. In: Workshop on Interconnection Networks for Parallel and Distributed Processing. West Layfaette, IN, pp. 11–19, April.
17. Kung, S. -Y. (1987). *VLSI Array Processors.* Englewood Cliffs, New Jersey: Prentice Hall.
18. Landman, P. E., Rabaey, J. M. (June 1995). Architectural Power Analysis: The Dual Bit Type Method. *IEEE Transactions on Very Large Integration (VLSI) Systems* 3(2):173–187.
19. Lengauer, C. (1993). Loop Parallelization in the Polytope Model. In: Best, E., ed. *CONCUR'93*, Lecture Notes in Computer Science 715. Springer-Verlag, pp. 398–416.
20. Marculescu, D., Marculescu, R., Pedram, M. (June 1996). Information Theoretic Measures for Power Analysis. *IEEE Transactions on Computer-Aided Design of Integrated Circuits and Systems* 15(6):599–610.
21. Moldovan, D. I. (January 1983). On the Design of Algorithms for VLSI Systolic Arrays. *Proceedings of the IEEE* 71:113–120.
22. Musoll, E., Cortadella, J. (1995). High-level Synthesis Techniques for Reducing the Activity of Functional Units. In: International Symposium on Low-Power Design, pp. 99–104, April 1995.

23. Musoll, E., Cortadella, J. (1995). Scheduling and Resource Binding for Low Power. In: Int. Symp. on System Synthesis, pp. 104–109.
24. PARO Design System Project. http://www-date.upb.de/research/paro/
25. Raghunathan, A., Jha, N. K. (1995). An ILP Formulation for Low Power based on Minimizing Switched Capacitance during Data Path Allocation. In: IEEE Symposium on Circuits and Systems, May.
26. Raghunathan, A., Jha, N .K., Dey, S. (1998). High-Level Power Analysis and Optimization. Norwell, Massachusetts: Kluwer Academic.
27. Synopsys, Inc., http://www.synopsys.com
28. Teich J. (1993). A Compiler for Application-Specific Processor Arrays. Ph.D. thesis, Institut für Mikroeletronik, Universität des Saarlandes, Saarbrücken, Germany.
29. Teich, J., Thiele, L., Zhang, L. (September 1997). Scheduling of Partitioned Regular Algorithms on Processor Arrays with Constrained Resources. *Journal of VLSI Signal Processing* 17(1):5–20.
30. Thiele, L. (1995). Resource Constrained Scheduling of Uniform Algorithms. *Journal of VLSI Signal Processing* 10:295–310.

7

Automatic Synthesis of Efficient Interfaces for Compiled Regular Architectures

Steven Derrien, Anne-Claire Guillou, Patrice Quinton, and Charles Wagner
Irisa, Campus de Beaulieu, Rennes, France

Tanguy Risset
LIP, ENS Lyon, France

I. INTRODUCTION

The technological road map for embedded system design foresees important changes in the design methodologies. *Interface protocol-based* and *platform-based designs* [30] are becoming mandatory because the complexity of the designs increases very fast, and time to market requirements are shorter and shorter. It is well-known, however, that technology improvements do not affect uniformly all elements of a chip: memories, buses, and power supply integration do not follow Moore's law. Thus, the main focus in hardware design has moved from performance and parallelism to other concerns such as the optimization of memory size, hierarchy and traffic, the elimination of the bus bandwidth bottleneck, and the minimization of power consumption. Power-aware design can be considered as today's most important problem, but with the emergence of *network on chip* [7], interfacing blocks of Intellectual Property (IP) efficiently becomes a major issue. This chapter addresses the latter problem.

Many designs rely on bus protocols to interface IP, for example AMBA or CORE-connected bus. This tendency is amplified by the spreading of the *platform-based design* methodology, which imposes the use of particular buses on the chip. But the gap between the design clock cycles and bus clock cycles increases and therefore, the speed at which data can be fed into a design becomes the major bottleneck for performance. Thus, a good design must be delivered with an interface protocol that uses the bus bandwidth efficiently. Such an interface is most often designed by hand, but clearly, automatic generation of interfaces must be considered. This research is mainly concerned with efficient bus protocol interface definition for highly parallel design. Note here that some attempts to plug Field Programmable Gate Arrays (FPGA) chips directly to memories have been made [16]. In such a framework, the interface synthesis may be very different from the one presented here. However, one part of the present work—how to extract interface information from high-level specifications—remains valid, even if it must be retargeted.

To be complete, the analysis of technological evolution should include trends in the electronic design automation (EDA) tools. Currently the software technology provides satisfying tools for logic synthesis (i.e., synthesis from RTL specification). But trends lead toward providing higher level design methodologies, using new languages (SystemC for instance) and new methodologies (UML specifications for instance). The research presented here belongs to a specific domain that aims at compiling loops to hardware [26, 13,10,2]. In this field, interfacing is also a major issue because applications operate on large data sets (usually streams in signal processing or images in multimedia), which must be efficiently brought onto the chip. Again, design automation is a major issue and the approach proposed in this chapter focus on automatic interface design for architectures compiled from loops nest specifications.

Based on the work done around the MMALPHA tool [10,8], we propose a solution to this problem for linear regular arrays. In this chapter, we introduce the concept of *application interface*, which can be seen as the application-dependent part of the interface for linear systolic arrays. As all highly pipelined designs share many properties, it is possible to define a *generic application interface* (i.e., an interface skeleton that is valid for all linear arrays and that can be easily parameterized for each application implemented). The experiments we report are oriented toward an FPGA platform, but the concepts can be used for interfacing IPs on a SoC, provided the IP has features that we will describe.

The underlying interface architecture that we consider is composed of a bus (with fixed bandwidth and throughput, possibly including a faster DMA mode) and of a FIFO that interconnects the bus and the application. The FIFO allows data to be buffered when an interruption occurs on the bus. The bus and the FIFO form what we call the *hardware interface*. On top of the hardware interface is built an *application interface* whose role is to rearrange the data between the hardware interface and the application.

This part of the interface is application-dependent and can be automatically produced by the same kind of tools that generate the hardware of the application.

This chapter is organized as follows. In Section II, we present briefly the MMALPHA environment. After a short description in Section III of the DLMS application that serves as an illustration, we explain in Section IV the model of our application architecture. Section V details the various elements of the interface we target. In Section VI, we describe how data transfers are structured in phases and patterns in order to obtain efficient communications. The generation of the interface, both software and hardware, is presented in Section VII. We then describe in Section VIII the use of our interface generator to automatically implement a DLMS filter on a FPGA board. Finally, we present in Section IX related work and we form conclusions in Section X.

II. THE MMALPHA SYSTEM

MMALPHA is a programming environment for transforming Alpha programs into parallel architectures based on research on automatic systolic synthesis [25]. Alpha [19,15] is a functional language that was designed to express both recurrence equation specifications and hardware descriptions of systolic arrays. MMALPHA was used to design applications pertaining to signal processing, multimedia, and bioinformatics [21,20].

MMALPHA targets both Asics and FPGAs. The final output of the compilation process is a synthesizable VHDL program that can be compiled with vendor tools. The interest of targeting FPGAs stems from the observation that regular designs are well-suited to the internal architecture of most FPGAs, which are made of locally connected configurable logic blocs (CLB).

The design flow with MMALPHA is illustrated in Fig. 1. A computation-intensive part of a program (usually a loop) is translated into Alpha and is then transformed using MMALPHA into a parallel hardware description at register transfer level (RTL). This RTL description comprises three parts:

The hardware part, written in VHDL represents the implementation of the loop on the FPGA.
The software part replaces the loop in the original program.
The hardware/software part handles the communication of data between the host and the FPGA.

Figure 2 shows an example of the Alpha program for a delayed least mean square adaptive filter that we shall describe in detail in Section III. Such a program, called a *system*, represents a set of recurrence equations

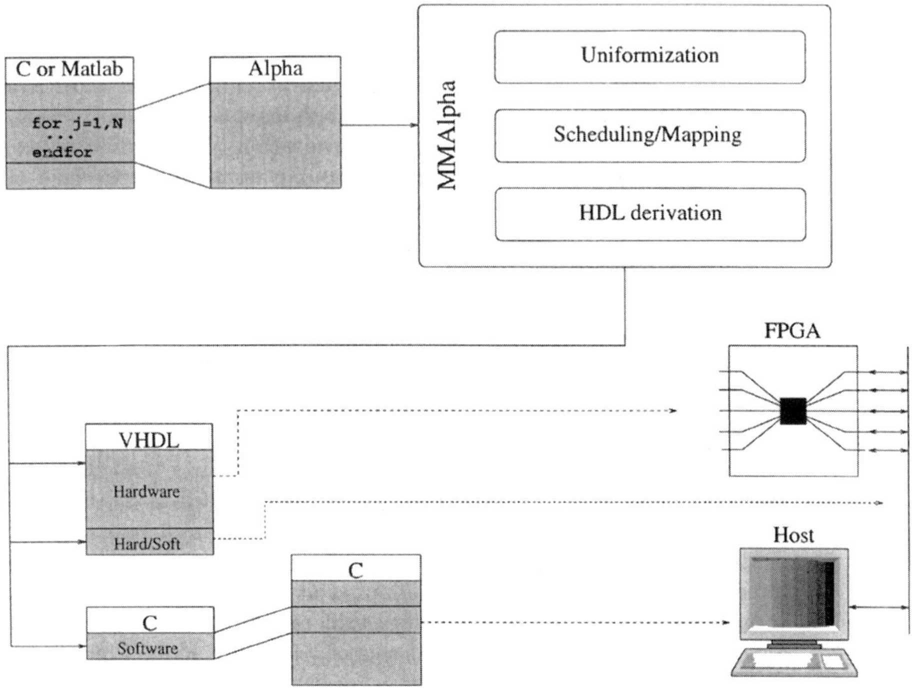

Figure 1 Design flow with MMALPHA. Square shaded boxes represent programs of various languages and round boxes represent transformations.

that allows iterative algorithms to be easily represented. Here, this system named `firr` is parameterized by the size N of the filter, the number of delays D, and the number of input samples M. It has two inputs, x and d, and produces one output y. Local variables W, Y, and E define the intermediate equations necessary to compute y.

Each equation of this program is an indexed expression whose index space is an integral polyhedron, that is to say, a set of integer coordinate points delimited by linear inequalities. Expressions are indexed by affine combinations of indices and of parameters. Operators (e.g., $+$, $*$, etc.) combine expressions as collections of data. Case expressions (see, for example, the definition of W) allow variables to be defined by different expressions on different polyhedral regions of the index space.

The synthesis process consists of applying a set of transformations, such as the ones mentioned in Fig. 1. Uniformization allows affine index expressions to be replaced by translations, thus removing potential broadcasts of the target architecture. Scheduling and mapping assign to each expres-

```
system firr: {N,M,D | 3<=N<=(D-1,M-D-1)}
  (x : {n | 1<=n<=M} of real;
   d : {n | N<=n<=M} of real)
returns
  (y : {n | N<=n<=M} of real);
var
  W : {n,i | N<=n<=M; 0<=i<=N-1} of real;
  Y : {n,i | N<=n<=M; -1<=i<=N-1} of real;
  E : {n | N<=n<=M} of real;
  mu : integer;
let
 mu[] = 0.5[];
 W[n,i] =
   case
    { | n<=N+D-1} : 0[];
    { | N+D<=n} : W[n-1,i] + E[n-D] * x[n-i-D];
   esac;
 Y[n,i] -
   case
    { | i=-1} : 0[];
    { | 0<=i} : Y[n,i-1] + mu[] * W[n,i] * x[n-i];
   esac;
  E[n] = d[n] - y[n];
  y[n] = Y[n,N-1];
tel;
```

Figure 2 Alpha program for a DLMS filter.

sion an execution time and processor. Finally, an RTL description is generated by translating the scheduled and placed Alpha code in an HDL language such as VHDL.

III. THE DLMS EXAMPLE

In this section, we detail the delayed least mean square algorithm (DLMS) for channel error correction in signal processing applications. This algorithm is used throughout this chapter to illustrate our method.

Least mean squares adaptive filters are commonly used in signal processing applications such as echo cancellation, system identification, speech coding, and channel equalization [11]. Unlike fixed coefficient finite impulse response (FIR) or infinite impulse response (IIR) digital filters, the coefficients of adaptive digital filters such as the LMS filter are adapted at each iteration to obtain better convergence properties. It is well known that recursive or adaptive digital filters are difficult to pipeline due to the presence of a feedback loop. However, it is possible to obtain a pipelined implementation by inserting delays in the recursive loop of the coefficient update part. The corresponding algorithm is thus called a *delayed least mean square* (DLMS) algorithm.

By assuming that the adaptive digital filter is an FIR filter whose impulse response is denoted by $w_i\,(n)$, the output signal $y\,(n)$ is given by

$$y(n) = \boldsymbol{x}^T(n)\boldsymbol{w}(n) = \sum_{i=0}^{N-1} x(n-i)w_i(n) \tag{1}$$

with $x(0) = 0$ and $\boldsymbol{w}(0) = 0$ (bold variables denotes N-vectors: $\boldsymbol{x}(n) = (\boldsymbol{x}(n - N - 1), \ldots, x(n))$ and $\boldsymbol{w}(n) = [w_0(n), \ldots, w_{N-1}(n)]$). The weight update equations of the DLMS [12] are given by

$$\boldsymbol{w}(n+1) = \boldsymbol{w}(n) + \mu\, e(n-D)\boldsymbol{x}(n-D) \tag{2}$$

$$e(n) = d(n) - y(n) \tag{3}$$

where $d(n)$ is the desired signal.

A possible VLSI implementation of the DLMS [12] is represented in Fig. 3. The RTL description of this architecture can be derived automatically from program of Fig. 2 with the MMALPHA software as shown in [10].

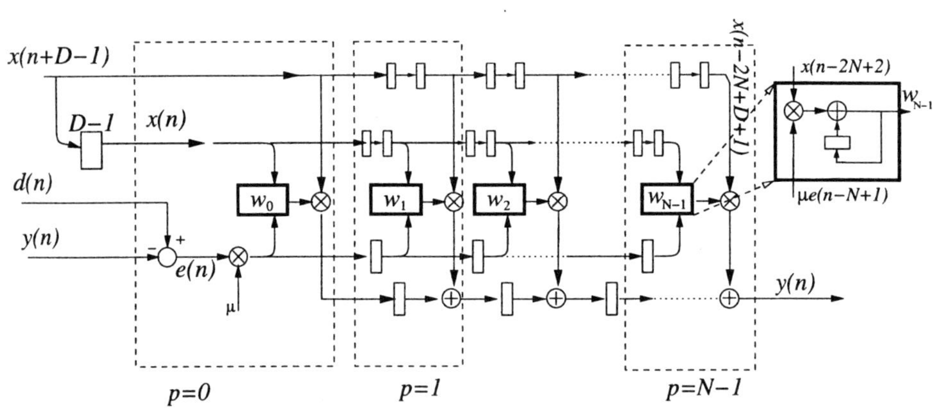

Figure 3 Snapshot (at $t = n - N + D$) of the architecture obtained for the DLMS after MMALPHA design process. N is the number of taps of the filter, D is the number of delays in the feedback loop.

At the end of the synthesis process, the architecture presented in Fig. 3 is expressed as an ALPHARD program, which is a subset of Alpha, and the mapping between the functional specification and the hardware implementation is obtained by means of the program of Fig. 4. More precisely, the input flow *x* and the coefficient vector *d* of the initial `firr` algorithm are mapped to new variables `x_mirr1`, `x_mirr2`, and the `d_mirr1`, and the architecture itself represented by an instantiation of another Alpha program called `firrModule` (by a `use` statement), which returns an output stream *Y1*. This stream is assigned to the output variable *y* of program `firr`. All inputs and outputs of `firrModule` are indexed by *t* and *p*, which represent respectively the time and the processor number to which these streams are assigned.

```
system firr :   {N,M,D | 3<=N<=(M-D-1,D-1)}
                (x : {n | 1<=n<=M} of integer[S,16];
                 d : {n | N<=n<=M} of integer[S,16])
       returns (y : {n | N<=n<=M} of integer[S,16]);
var
  d_mirr1 : {t,p | D<=t<=-N+M; p=0} of integer[S,16];
  y_mirr1 : {t,p | D<=t<=-N+M; p=0} of integer[S,16];
  x_mirr1 : {t,p | -N+D+1<=t<=-N+M; p=0} of integer[S,16];
  x_mirr2 : {t,p | -N+2<=t<=-N+M+1; p=0} of integer[S,16];
  Y1 : {t,p | N<=t<=M; p=N-1} of integer[S,16];
let
  y_mirr1[t,p] = y[t+N-D];
  d_mirr1[t,p] = d[t+N-D];
  x_mirr2[t,p] = x[t+N-1];
  x_mirr1[t,p] = x[t+N-D];
  y[n] = Y1[n,N-1];
  use  firrModule[N,M,D] (d_mirr1, y_mirr1, x_mirr1, x_mirr2)
                 returns  (Y1) ;
tel;
```

Figure 4 The part of the ALPHARD program (`firr` system, also called ALPHARD *interface*) which maps the functional specification (input *x* and *d*, result *y*) to the architecture (`firrModule` system also called ALPHARD *module*). This program has three parameters: *N* is the number of taps of the filter, *D* is the number of delays along the feedback loop, and *M* is the number of input samples of the filter (for simulation purposes). This system contains the information about the date and place where data should be entered (here for example, input in the first processor: $p = 0$, and output in the last processor: $p = N - 1$).

IV. APPLICATION ARCHITECTURE MODEL

This section models the interface we want to synthesize. We first state the assumptions that we make regarding the type of application hardware that we want to interface. Then we detail the information that is needed for interfacing correctly the application architecture (e.g. `firrModule` in the DLMS application) with its host architecture. Then we abstract this architecture by a number of features that will constitute the input to interface generation.

A. Assumptions

Our assumptions regarding the application architecture concern four aspects: a virtual clock, the model of linear array, inputs and outputs, and the bit width of streams.

1. Virtual Clock

The application architecture is a *globally synchronous digital circuit* in which all registers are controlled by a *common virtual clock*. This virtual clock regulates the operation of the architecture, and can be frozen if, for instance, the host is not ready to send input data or to read output data. This assumption significantly reduces the control complexity inside the interface. Indeed, the designer can assume that as soon as the clock of the architecture is running, input data arrive as needed, and output data are captured by the bus when they are produced. In our experimental designs, which targets FPGA chips, the virtual clock is naturally implemented using the *clock enable* signal of the FPGA. We also assume that the operation of the architecture begins on a `start` signal.

From now on, we call *virtual date* of a computation the number of virtual clock cycles elapsed between the computation and the starting time of the algorithm.

2. Linear Arrays

The architecture must be a linear (1-dimensional) array. Inputs and outputs are continuous streams of data, which means that all input or output streams have a virtual starting date and a virtual ending date, and no interruption occurs between these dates. It should be noticed that this assumption prevents one from interfacing *partitioned arrays* [4], and *2-dimensional arrays*. In partitioned arrays, data arrive in *burst* mode (i.e., uninterrupted streams of data separated by long empty periods), and do not meet our assumption of continuous streams. However, the interface proposed in this chapter could be easily extended to cover this case because these bursts of data are known statically [5]. In MMALPHA, architectures can be generated for any 2-dimensional regular array, but in practice, interfacing a 2-dimensional array requires a more complex architecture, since more than one data must be provided during one virtual clock cycle. This problem is beyond the scope of this chapter.

3. Inputs and Outputs

Any given input stream (resp. output stream) must arrive in (resp. leave) a given fixed processor, called *connection processor* of the stream. Notice that this processor may be different for each input or output stream. This assumption is most often met by the type of architecture that we are dealing with, and is easy to enforce if not.

4. Width of Streams

We assume that the bit width of the streams is a divisor of the data bus width. As the bus is usually 32- or 64-bit wide, the bit width of variables must be a power of two (if it is not, the protocol will choose the smallest power of two greater than the actual bit width, thus reducing the efficiency of the interface).

B. Information Needed for the Interface

If the above assumptions are met, then the interface of the architecture can be determined from the following information.

- The number of input and output streams. In our example, the `firrModule` system has four inputs (`d_mirr1`, `y_mirr1`, `x_mirr1`, `x_mirr2`) and one output (`Y1`).
- The name, bit width, connection processor, virtual starting and ending time of each stream. For instance input stream `x_mirr1` is 16-bit wide and is input in processor $p = 0$. The starting time is $t = -N + D + 1$ and the ending time is $t = -N + M + 1$.

All this information can be extracted from the ALPHARD interface shown in Fig. 4 (see, for example, the declaration of the `x_mirr1` variable in Fig. 4).

V. INTERFACE MODEL

The previous section described *what* we want to interface as well as the information needed to define this interface. We now explain in more detail our interface model. Fig. 5 presents a typical interface architecture, in the case of the FPGA Spyder board [28]. The application hardware, here a

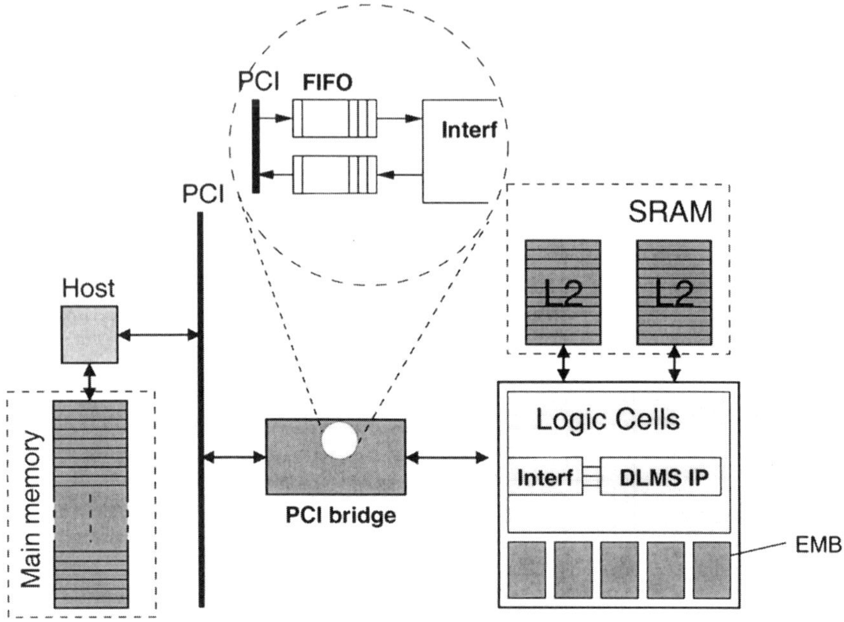

Figure 5 Standard architecture of the interconnection between a board and the host (taken from a Spyder board [28]), together with a logical view of the FIFO mechanism used between the bus and the application architecture.

DLMS filter circuit, is mapped on the FPGA. The host and the DLMS are interconnected by means of a PCI bus. A PCI bridge consisting of FIFO allows for a smooth synchronization between the PCI bus and the application hardware.

In this section, we detail the elements of this interface: the low-level interface, the application interface, and the software and hardware parts of the application interface.

A. Low-Level Interface and Application Interface

From now on, the interface is logically divided into two parts: the *low-level interface* and the *application interface*.

The low-level interface behaves logically as a FIFO. A parameter of the low-level interface is the FIFO *bit width*, which we assume to be a power of 2. Notice that the implementation of the low-level interface can take different forms, depending on the target platform. In the case shown in Fig. 5, the FIFOs of the low-level interface are implemented using the memory of the Spyder board. Most commercially available FPGA boards provide a similar low-level interface [1,3].

The *application interface* is the part of the interface that sends input data to (resp. gets output data from) the application IP from (resp. to) the host in order to implement a correct execution of the algorithm. The application interface is naturally divided into the *input interface*, which sends data to the array, and the *output interface*, which receives data from the array. These two parts are very similar, and from now on we only deal with the input interface.

B. Software and Hardware Parts

The application interface is naturally divided into a *software part* and a *hardware part*. The software part is composed of a program that sends data to the FIFO. The hardware part is more complex: it is the demultiplexing system that gets the words out of the FIFO and sends them into the array. Of course, these two parts must be compatible (i.e., the data should be taken by the hardware part in the same order as they are produced by the software part). This is why we propose to generate the software and the hardware simultaneously from the interface ALPHARD program shown in Fig. 4.

VI. STRUCTURING STREAMS

To generate the application interface, we structure the input and output streams using two notions: *phases* and *patterns*.

A. Phases

A phase is a sequence of successive virtual clock ticks during which all inputs and outputs of the architecture are the same. For instance, one can see in the program of Fig. 4 that between clock cycles $t = -N + 2 = -8$ and $t = -N + D = 2$, only the `x_mirr2` stream enters the array. Hence the period of time

$$\phi_0 = \{t \mid -N + 2 \leq t \leq -N + D\} \tag{4}$$

is a phase.

Finding the phases of the program of Fig. 4 can be done with the help of elementary computations on polyhedra. The algorithm is the following.

1. Compute the *time domain* of each input and output variable (i.e., the period of time during which this variable is alive). This is obtained by projecting the space-time domains (see Fig. 4) of these

variables on the time index t. Since we assume that inputs and outputs consist of uninterrupted streams of data, for each input or output variable v, the result is a set of intervals of the form $\{t \mid l_v \leq t \leq u_v\}$. Note that projections of polyhedra can be computed using the Polylib library [29].

2. Sort the lower and upper bounds of these time domains. The result is an ordered list of dates $(b_i)_{i=1,\ldots}$ such that $b_i \leq b_{i+1}$.
3. The phases are the intervals $\{t \mid b_i \leq t < b_{i+1}\}$ for all i.

For instance, after projecting the domains of variables `x_mirr2` and `x_mirr1` in the program of Fig. 4, we obtain the time domains

$$\{t \mid -N+2 \leq t \leq -N+M+1\}$$

and

$$\{t \mid -N+D+1 \leq t \leq -N+M\}$$

As the calculation of all other variables starts later, the first phase is

$$\phi_0 = \{t \mid -N+2 \leq t \leq -N+D\}$$

In practice, we constrain phases to meet an additional requirement: the number of bits that are sent during a phase must be a multiple of the bus width. Indeed, during a phase, data is sent to the FIFO word by word, each word having the size of the FIFO width. For instance the `x_mirr2` variable being 16-bit wide, and assuming the FIFO to be 32-bit wide, two `x_mirr2` data can be placed in one FIFO word (remember that we assume the bit width of all streams to be a divisor of the FIFO width). Thus, we require the length of a phase to be a multiple of the $\frac{\text{FIFO width}}{\text{data width}}$ ratio. If this condition is not fulfilled, one can always split the phase in two subphases, with one meeting this condition and the other having less bits than one FIFO word.

This requirement forces us to fix the value of the parameters at this stage. In our example (32-bit wide FIFO and 16-bit wide variable), the ratio $\frac{\text{FIFO width}}{\text{data width}}$ is 2, and phase ϕ_0 in Eq. (4) contains $D-1$ virtual clock cycles. If D is even, this interval must be divided into two phases: phase $\phi_1 = \{t \mid -N+2 \leq t \leq -N+D-1\}$ and phase $\phi_2 = \{t \mid t = -N+D\}$. This situation is illustrated in Fig. 6, which shows the phases of the program of Fig. 4, for $N = 10$, $D = 12$, and $M = 100$. We can see that $\phi_1 = \{t \mid -8 \leq t \leq 1\}$ and $\phi_2 = \{t \mid t = 2\}$.

B. Patterns

Inside each phase, a *pattern* describes the order in which data are sent to the FIFO. A pattern is repeated cyclically during one phase. Recall that each

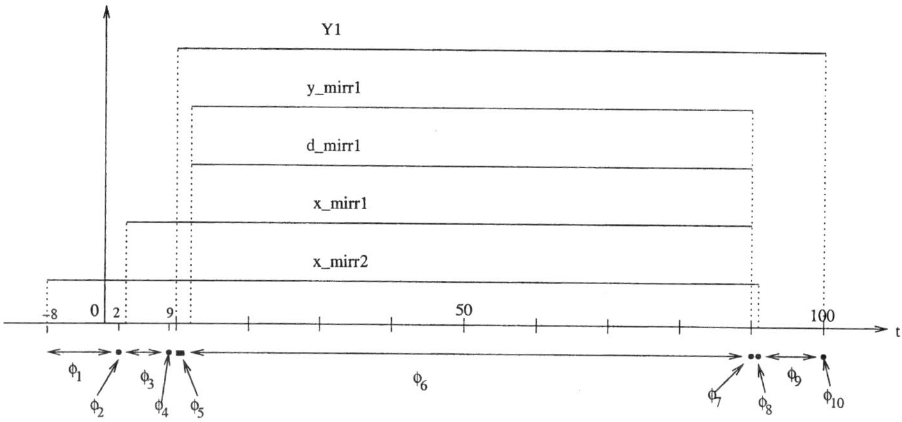

Figure 6 The phases of the program of Fig. 4 for parameter values $N = 10$, $D = 12$, and $M = 100$.

FIFO word contains only values related to one variable. For instance, during phase ϕ_3 of Fig. 6, one can choose to fill a FIFO word with two `x_mirr1` data and then the next FIFO word with two `x_mirr2` data and repeat this pattern three times so that six data of each stream are sent. A pattern is therefore simply an ordered list of variables that will be sent through the bus to the application interface. In the case of phase ϕ_3, the pattern is (`x_mirr1, x_mirr2`). In this particular case, the pattern is simple since `x_mirr1` and `x_mirr2` have the same bit width.

In general, the choice of a pattern must be done in such a way that deadlock situations may not occur due to a lack of memorization resources in the interface.

Let us assume that the IP to be interfaced has two input signals A and B during a particular phase. Denote respectively BW_A and BW_B the bit width of A and B, and assume that $BW_A = 16$ and $BW_B = 2$. Let $BW = 32$ be the bus width. Denote respectively ND_A and ND_B the *number of data* contained in one bus word for each variable. Then $ND_A = \frac{BW}{BW_A} = 2$ and $ND_B = \frac{BW}{BW_B} = 16$. Assume that the interface has only one BW bits word of memory available to store each variable (this assumption is reasonable in order to keep the interface simple).

To start the computations of the phase, the bus must send to the FIFO one bus word containing A values (i.e., 2 values) followed by one bus word containing B values (i.e., 16 values). These values will feed the IP architecture during two virtual clock cycles. If the next word sent by the bus was composed of B values, then the IP would be blocked, as the interface would not be able to store this second word before getting the next A value. Thus, the next data on the bus *must* be an A as well as the next six following bus words. In the end, we have the *pattern* BAAAAAAAA, which must be cyclically repeated during the whole phase. Note that the pattern ABAAAAAAA is also valid, and the solution is therefore not unique.

Two methods can be used to compute a dealock-free pattern. First, one can *enumerate* the patterns, which amount to simulating the process described in the previous paragraph. This method may be effective, as the pattern length is less than 64, and as we assumed the bit width of variables to be less than 32. Second, a formula describing valid patterns can be obtained by the following algorithm, which we call the *parallel pattern* method.

Assume that each variable has its own bus, instead of sharing a common bus between the interface and the IP. Then, a *parallel pattern* is a list of dates when each variable can be sent to its bus. [In such a model (in the previous example), an A word and a B word could be sent in parallel.] A method to generate deadlock-free patterns consist of finding a parallel pattern and then in serializing it without introducing deadlock.

To find parallel patterns, the method is as follows.

1. Compute the *parallel pattern length* PPL. This length is the smallest number of data that can be sent cyclically to the bus without deadlock. It can be proved that this value is

$$\mathrm{PPL} = \frac{\max_V \mathrm{ND}_V}{\min_V \mathrm{ND}_V}$$

   where ND_V is the number of data contained in one bus word for variable V. For instance in the above example, PPL = 16/2 = 8. Indeed, A is repeated 8 times in the valid patterns mentioned for the example, and this value is always an integer because we assume that the bit width are powers of 2. The *pattern cycle length* (PCL), that is to say, the number of virtual clock cycles executed by the IP during one pattern, is therefore

$$\mathrm{PCL} = \mathrm{PPL} \times \min_V \mathrm{ND}_V = \max_V \mathrm{ND}_V$$

2. For each variable V, $\frac{\mathrm{PCL}}{\mathrm{ND}_V}$ bus words must be sent during the pattern of length PPL. It is easy to see that these words must be regularly spaced. The time delay between two consecutive bus words of variable U is thus

$$\frac{\mathrm{PPL}}{\mathrm{PCL}/\mathrm{ND}_U} = \frac{\mathrm{ND}_U}{\min_V \mathrm{ND}_V}$$

and these data must be sent at time steps:

$$0, \frac{ND_U}{\min_V(ND_V)}, 2 \times \frac{ND_U}{\min_V(ND_V)}, \ldots, \left(\frac{PCL}{ND_U} - 1\right) \times \frac{ND_U}{\min_V(ND_V)}$$

Using this method, we obtain a parallel order for sending the data to the bus. In our example, we have $PPL = 8, \frac{ND_A}{\min(ND_A, ND_B)} = 1$ and $\frac{ND_B}{\min(ND_A, ND_B)} = 8$, so that the parallel order is

Time steps	0	1	2	3	4	5	6	7
Dates for A	A	A	A	A	A	A	A	A
Dates for B	B	|	|	|	|	|	|	|

where | means no data is sent.

From a parallel pattern, one easily obtains a sequential pattern by interleaving the branches of the parallel pattern in such a way that the partial order imposed by the parallel pattern is satisfied. In our example, the initial A and B may be sent in any order, which leads to one of the two solutions ABAAAAAAA or BAAAAAAAA as proposed previously. This also shows that these are the only valid solutions.

VII. GENERATING THE INTERFACE

As seen in the previous section, the *interface model* is abstracted by the description of its phases and, for each phase, of its patterns. We now explain how the software and the hardware part of the interface are generated from these informations.

A. Generating the Software Part

The software part of the application interface is a C program that the host runs in order to send data to the input FIFO or read data from the output FIFO through the PCI bus. The exact behavior of the C code is completely specified by the phase and pattern information. However, during the generation of the C code, many details must be handled carefully, like the conversion between the ALPHA variable indexing and the implementation in memory (array starting at index 0) for instance. All technical difficulties have been solved in the ALPHA to C compiler and are described elsewhere [24].

The software part is therefore produced by a C program, which was obtained by retargeting the Alpha to C compiler.

Fig. 7 shows a part of the C program that was generated to perform inputs and outputs of phases ϕ_6, ϕ_7, and ϕ_8 of Fig. 6. Each call to `WriteFifo` or `ReadFifo` activates a function of a low-level communication library. Note that the *t* index corresponds precisely to the *t* index of the ALPHA program of Fig. 4 as well as to the *t* coordinate of Fig. 6, thus the designer can easily understand what is going on during the execution of the software interface.

Notice that in this C program, inputs and outputs are interleaved, hence the FIFO is not really needed (a FIFO of size one would be sufficient). Inserting a FIFO between the architecture and the host is however useful in order to interface the IP to a more general system. Instead of single C program, another possibility would be to have inputs and outputs handled by two different processes running in parallel on the host.

```
/* phase 6 : from 12 to 89 with variables:
      {d_mirr1, y_mirr1, Y1, x_mirr1, x_mirr2}*/
for (t = 12; t <= 89; t = t + 2) {
   WriteFifo( (int *)(_d_mirr1 + (t -12)));
   WriteFifo( (int *)(_y_mirr1 + (t -12)));
   ReadFifo( (int *)(_Y1 + (t -10)));
   WriteFifo( (int *)(_x_mirr1 + (t -3)));
   WriteFifo( (int *)(_x_mirr2 + (t+8)));
}
/* phase 7 : from  90 to 90 with variables:
      {d_mirr1, y_mirr1, Y1, x_mirr1, x_mirr2} (1 data sent)*/
t = 90; {
   WriteFifo( (int *)(_d_mirr1 + (t -12)));
   WriteFifo( (int *)(_y_mirr1 + (t -12)));
   ReadFifo( (int *)(_Y1 + (t -10)));
   WriteFifo( (int *)(_x_mirr1 + (t -3)));
   WriteFifo( (int *)(_x_mirr2 + (t+8)));
}
/* phase 8 : from  91 to 91 with variables:
     {Y1, x_mirr2} (1 data sent)*/
t = 91; {
   ReadFifo( (int *)(_Y1 + (t -10)));
   WriteFifo( (int *)(_x_mirr2 + (t+8)));
}
```

Figure 7 C code generated for phases ϕ_6, ϕ_7, and ϕ_8 of Fig. 6 (for a 32-bit wide FIFO). Array `_d_mirr1` stores the values of the `d_mirr1` variable of the ALPHA program.

B. Generating the Hardware Part: Principles

The hardware part is generated in VHDL and is synthesized for the FPGA chip. The architecture of the hardware part is illustrated in Fig. 8.

The hardware part is divided into the input interface and the output interface, which are almost symmetrical. The main difference is that the interface is started with a `start` signal (set up by the user, here indicating the virtual date $t = -8$), while the output interface is started by a `start_out` signal set up automatically (here at virtual date $t = 10$).

We detail the organization of the input interface represented in Fig. 8. Consider an input, say input I1, of the architecture, and let W_{I1} be the bit width of I1. This input is connected to a `load32` component which is parameterized by the bit width W_{I1} of I1. The `load32` component contains a shift register that allows 32 bits to be read in parallel from the FIFO and W_{I1} bits to be output during $32/W_{I1}$ clock cycles.

Each `load32` component is connected to the `Input_Interface` component. The `Input_Interface` component receives data from the FIFO and stores them in the appropriate `load32` shift register.

The control of this architecture is provided by a hierarchical two-level finite state machine. The states of first level are the phases of the interface, and the states of the second level are the variable names which define the patterns inside a phase. Switching from one phase to another is done by counting the number of elapsed virtual clock cycles (for example, we see on Fig. 6 that phase ϕ_1 must last 10 virtual clock cycles). An efficient control of the `load32` shift register allows the loading of a new FIFO word to be overlapped with the output of the last data word to the application architecture. Thus, provided that the FPGA clock frequency is high enough, the array is fed at the throughput allowed by the bus. Usually, the FPGA clock is fast enough to cope with the bus bandwidth, since the application design

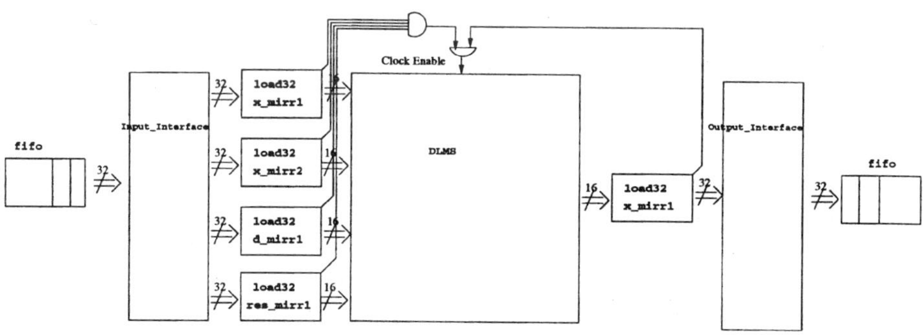

Figure 8 Hardware part of the application interface for the DLMS.

is highly pipelined. However, one can imagine extreme situations where this is not the case. For example, if the input data was only 2 bit wide, the FPGA clock frequency would have to be 16 times the bus clock frequency, which is not very realistic. Note that the operation of the interface and of the application architecture is frozen when the FIFO is empty.

C. Efficient Synthesis of the Hardware Part

The originality of our approach lies in the fact that we designed a *family* of interfaces that can be parameterized for different IPs provided they respect the model of Section IV. Most of the produced VHDL code is generic (i.e., identical for all IP, and, for each IP, a *configuration file* which drives the control of the IP is generated). In this section, we show how to efficiently implement this genericity in VHDL. Indeed, the performance of the final IP is greatly influenced by the quality of the interface description.

The VHDL program describes a finite state machine which is identical for all application architectures meeting the assumption of section IV (see Appendix IX). The information on one phase (i.e., pattern, duration, active variables, etc.) is stored in a VHDL `record`. The finite state machine uses an `array of records` whose size is the number of phases (here 10 phases). With this implementation, one can gather all the application-dependent information in one configuration file. This file is built by extracting information from the interface program of Fig. 4 using the *same* function for finding the phases and pattern as the one used during the C code generation for the software part. Thus, and this is a very important feature of our methodology, compatibility between the software part and the hardware part is very easy to ensure as they both derive from the same source code (program of Fig. 4 with the same function used for common information). This way of representing the finite state machine simplifies the automatic generation of the VHDL interface program for all applications. Moreover it guarantees a very efficient synthesis with the vendor's tools.

VIII. EXPERIMENTS

The automatic interface generation was implemented and tested with a Spyder-X2 PCI board [28] based on a Xilinx Virtex 800 device (as shown in Fig. 5). In this architecture the host processor communicates with the FPGA through a PCI interface using memory-mapped or DMA transfers. The observed bandwidth between the host CPU and the FPGA is 8 MB/s at most.

The VHDL for the DLMS was synthesized automatically from an ALPHA specification down to ALPHARD (see [10]). An important issue during the design process was functional simulation. Thanks to the flexibility of the MMALPHA environment, we were able to use the same data from high-level simulation down to the detailed VHDL simulation, thus speeding up the design time and allowing fast back annotation from hardware simulation to the high-level synthesis process.

The application interface was generated automatically from the ALPHARD interface for the DLMS algorithm (Fig. 4) for the following values of the parameters: $N = 10$, $D = 12$, and $M = 100$. The width of the interface FIFO was 32 bits. The synthesis was realized with the SYNPLIFY software [27]. Table 1 gives the number of look-up tables necessary for the synthesis of the DLMS alone, of the DLMS and the interface without the FIFO, and of the total design. The clock cycle, as estimated by the synthesis tool, is also given. Finally, the throughput of the interface is evaluated from the cycle time and the number of data that is produced by the architecture during each cycle. More precisely, the DLMS produces one, 16-bit `y_mirr` value during each cycle. The table also shows the maximum throughput of the PCI bus of the board, as observed for several designs.

One may draw some conclusions from this table.

First, the hardware interface is not a limiting factor of speed for this design. Indeed, the clock cycle is increased only by 1 ns by the FIFO.

Second, the PCI bus is clearly the limiting factor of the interface: there is a factor of 8 in the best case between the bandwidth of the bus and the bandwidth that could be achieved by the design. As expected, the design of such a high-performance device is therefore limited by the communication with the host.

Table 1 Result of the Interface Generation for the DLMS Algorithm

Design	Number of LUT	Clock cycle (ns)	Throughput (MB/s)
DLMS	5938 (31%)	30	66.67 MB/s
DLMS + interface	6501 (34%)	30	66.67 MB/s
DLMS + interface + Fifos	6928 (36%)	31	64.5 MB/s
PCI bus (Max)	—	—	8 MB/s

Parameters: $N = 10$, $D = 12$, and $M = 100$. This table gives the number of look-up tables (LUT) occupied by the design, in the Virtex XCV800 chip (the percentage of total LUTs used in a Virtex XCV800 is given between parentheses), the clock cycle estimated by the synthesis tools, and the (one-way) throughput of the interface. The maximum observed throughput of the PCI bus of the Spyder board is given for comparison.

Notice that this interface could have been optimized since the `x_mirr1` and `x_mirr2` streams are a time shift of one another, thus only one of the two streams needed to be sent through the bus. Moreover, the `y_mirr1` stream should, in practice, be taken at the output of the architecture and not sent from the host (the values sent were obtained during simulations). We kept this nonoptimized implementation for our experiments in order to validate the interface protocol on a nontrivial example.

IX. RELATED WORK AND DISCUSSION

Most of the research on FPGA design focuses on the efficiency of the design itself rather than the efficiency of the interface of the design. Tests are usually made with data already in the on-board memory or with data arriving directly on board via an analog/digital converter. Among others, references [14,9] describe manually written interfaces.

Approaches dealing with automatic generation of interfaces can be classified into two categories: co-design methods and high-level synthesis methods.

Co-design methods deal with very general types of applications and, therefore, communication is usually implemented using a high-level synchronization mechanism that leads to complex protocols (for instance, remote procedure call in CoWare [6]). Many FPGA compilation projects rely on control-dominated models like Petri nets [31,18], or communicating sequential processes [22]. The advent of real-time operating systems (RTOs) can also be a solution but efficiency will probably be degraded.

In high-level synthesis methods, attempts have been made to automate interface generation. These attempts, as in the work presented here, restrict the types of interfaced architectures. In Pico [26], the interface problem is solved at run time by a bus arbiter, but since the target architecture has only a small number of processors, the efficiency of the interface is not the major issue. Artemis [23,13] relied on the Spade methodology [17] to implement efficiently communicating Kahn process networks on buses.

The methodology that we propose for the interface generation presents several novelties. First, dynamic control is as low as possible. Indeed, it is restricted to ensuring a correct behavior when an external event such as a bus interruption occurs. Therefore, the efficiency of the interface can be statically predicted. Second, the design is safe because the interface is compiled *together with the architecture*, and moreover, the hardware and software parts are derived from *the same* ALPHARD program and using *the same tool*. Third, our method provides the user with simulation facilities that are available as part of the MMALPHA tool. Finally, our model is generic and it is not only an implementation dedicated to the MMALPHA design flow, but can be applied to any compiled architecture that meets the characteristics of our model.

X. CONCLUSION

In this chapter, we presented a tool for synthesizing interfaces for linear regular architectures. Our model of interface is bus-based and can cope with interruptions in the flow of data arriving from the host. Our interface is application-dependent, and is generated automatically from a high-level description of the application, as obtained using the MMALPHA tool. Both the software part and the hardware part of the interface are generated from the same description, using the same set of tools, therefore ensuring these parts to be coherent. The synthesis of the interface structures the data communications into phases and patterns, and generates a C program to be run on the host, and a VHDL program to be synthesized on the hardware platform. We have tested this interface generator on a DLMS algorithm automatically compiled on a Spyder FPGA board and we have shown that the interface allows high performances to be reached for this application.

XI. APPENDIX: VHDL CODE OF THE APPLICATION-DEPENDENT INTERFACE

This part of the program stores into an array (`Input_PhaseTable`) the information about input phases (i.e., phases where input variables are concerned). `Input_ToPhase` is a function that fills the record with the input values.

```
constant Input_Nb_IO_Port: integer := 4;

-- Total number of input phases
constant Input_NbPhase: integer :=9;

-- Maximum number of variables in pattern
constant Input_MaxStep: integer := 18;

-- Maximum size of a pattern
constant Input_MaxLength: integer := 5;

-- Max Number of virtual clock cycles in a phase
constant Input_MaxCount: integer := 79;

subtype Input_TIO_select is
    std_logic_vector(Input_Nb_IO_Port-1 downto 0);
```

```
subtype Input_TStep is integer range 0 to Input_MaxStep;
subtype Input_TLength is integer range 0 to Input_
  MaxLength;
subtype Input_TCount is integer range 0 to Input_MaxCount;

-- Record used to store all Phase Information
Type Input_TPhaseInfo is record
    -- offset of the pattern in the pattern table
    Offset          : Input_TStep;
    -- Length of the pattern of the phase
    PatternLength   : Input_TLength;
    -- Which ports are active in the phase
    ActivePorts     : Input_TIO_select;
    -- how may virtual clock cycles in the phase
    NbArrayCycles   : Input_TCount;
end record;

type Input_TPhaseTable is array (0 to Input_NbPhase)
  of Input_TPhaseInfo;

constant Input_PhaseTable : Input_TPhaseTable :=
(
--     Offset PatternLength PhaseLength NBCycle Activity
    0 => Input_ToPhase (0, 1, 4, (others => '0')),
    1 => Input_ToPhase (1, 1, 10, Input_Portx_mirr2),
    2 => Input_ToPhase (2, 1, 1, Input_Portx_mirr2),
    3 => Input_ToPhase (3, 2, 6, Input_Portx_mirr1 or
                        Input_Portx_mirr2),
    4 => Input_ToPhase (5, 2, 1, Input_Portx_mirr1 or
                        Input_Portx_mirr2),
    5 => Input_ToPhase (7, 2, 2, Input_Portx_mirr1 or
                        Input_Portx_mirr2),
    6 => Input_ToPhase (9, 4, 78, Input_Portd_mirr1 or
                        Input_Porty_mirr1 or
                        Input_Portx_mirr1 or Input_
                          Portx_mirr2),
    7 => Input_ToPhase (13, 4, 1, Input_Portd_mirr1 or
                        Input_Porty_mirr1 or
                        Input_Portx_mirr1 or Input_
                          Portx_mirr2),
    8 => Input_ToPhase (17, 1, 1, Input_Portx_mirr2),
    9 => Input_ToPhase (0, 1, 4, (others=>'0'))
);
```

REFERENCES

1. Annapolis Micro System, Inc. WildStar Datasheet, 2001.
2. Catthoor, F., Danckaert, K., Kulkarni, C., Omnes T. (2000). Data Transfer and Storage Architecture Issues and Exploration in Multimedia Processors. In: *Programmable Digital Signal Processors: Architecture, Programming, and Applications*. New York: Marcel Dekker Inc.
3. Celoxica Ltd. RC1000 Datasheet, 2001.
4. Darte, A. (1991). Regular Partitioning for Synthesizing Fixed-Size Systotic Arrays. *Integration, The VLSI Journal* 12:293–304.
5. Darte, A., Rau, B., Vivien, F., Schreiber, R. (1999). A Constructive Solution to Juggling Problem in Systolic Array Synthesis. Technical Report 1999-15, Laboratoire de l'informatique du parallélisme.
6. De Man, H., Bolsens, I., Lin, B., Van Rompaey, K., Vercauteren, S., Verkest, D. (1997). Hardware and Software Codesign of Digital Telecommunication System. *Proceedings of the IEEE* 85(3):391–418.
7. De Micheli, G. (2002). Network on Chip: a new Paradigm for System on Chip Design. *Design Automation and Test in Europe*. Paris: IEEE Computer Society Press, pp. 418–420.
8. Derrien, S., Risset, T. (2000). Interfacing Compiled FPGA Programs: the MMALPHA Approach. In: Arabnia, A., ed. *PDPTA 2000: Second International Workshop on Engineering of Reconfigurable Hardware/Software Objects*. CSREA Press, June.
9. Frigo, J., Gokhale, M., Lavenier, D. (2001). Evaluation of the Streams C C-to-FPGA Compiler: an Applications Perspective. In: *Ninth international symposium on Field programmable gate arrays*. ACM Press, pp. 134–140.
10. Guillou, A.C., Quinton, P., Risset, T., Massicotte, D. High-Level Design of Digital Filters in Mobile Communications. DATE Design Contest 2001, March 2001. Second place, available at http://www.irisa.fr/bibli/publi/pi/2001/1405/1405.html
11. Haykin, S. (1996). *Adaptive Filter Theory*. 3rd ed. Prentice-Hall information and system sciences series. Upper Saddle River, NJ 07458, USA: Prentice-Hall.
12. Katsushige, M., Kiyoshi, N., Hitoshi, K. (1999). Pipilined LMS Adaptive Filter Using a New Look-Ahead Transformation. *IEEE Transactions on Circuits and Systems,* 46:51–55, January.
13. Kienhuis, B., Rijpkema, E., Deprettere, E.F. (2000). Compaan: Deriving Process Networks from Matlab for Embedded Signal Processing Architectures. In: *8th International Workshop on Hardware/Software Codesign (CODES'2000)*.
14. Lavenier, D., Quinton, P. (1996). SAMBA: Systolic Accelerator for Molecular Biological Application. Technical Report 988, Irisa, March.
15. Le Moenner, P., Perraudeau, L., Rajopadhye, S., Risset, T. (May 1996.). Generating Regular Arithmetic Circuits with AlpHard. In: *Massively Parallel Computing Systems (MPCS'96)*.
16. Leong, P.H.W., Leong, M.P., Cheung, O.Y.H., Tung, T., Kwok, C.M., Wong, M.Y., Le Pilchard, K.H. (2001). A Reconfigurable Computing Platform with

Memory Slot Interface. In: *Symposium on Field-Programmable Custom Computing Machines (FCCM)*. California: IEEE Computer Society Press.

17. Lieverse, P., Van der Wolf, P., Deprettere, F., Vissers, K. (2001). A Methodology for Architecture Exploration of Heterogeneous Signal Processing Systems. *Journal of VLSI Signal Processing for Signal, Image and Video Technology* 29(3):197–207. Special issue on SiPS'99.
18. Lin, B., Vercauteren, S. (1994). Synthesis of Concurrent System Interface Modules with Automatic Protocol Conversion Generation. In: *International Conference on Computer-Aided Design (ICCAD)*, pp. 101–109.
19. Mauras, C. (1989). *Alpha: un langage équationnel pour la conception et la programmation d'architectures parallèles synchrones*. Thèse de doctorat, Ifsic, Université de Rennes 1, December.
20. Mémin, E., Risset, T. (1999). Full Alternate Jacobi Minimization and VLSI Derivation of Hardware for Motion Estimation. In: *Int. Workshop on Parallel Image Processing and Analysis, IWPIPA'99*. India: Madras, Jan.
21. Mozipo, A., Massicotte, D., Quinton, P., Risset, T. (1999). A Parallel Architecture for Adaptative Channel Equalization based On Kalman Filter Using MMALPHA. *1999 IEEE Canadian Conference on Electrical & Computer Engineering*, pp. 554–559, May.
22. Page, I. (1996). Constructing Hardware-Software Systems from a Single Description. *Journal of VLSI Signal Processing* 12:87–107.
23. Pimentel, A. D., Hertzberger, L. O., Lieverse, P., Van Der Wolf, P., Deprettere E. F. (2001). Exploring Embedded-Systems Architectures with Artemis. *IEEE Computer* 34(11):57–63.
24. Quilleré, F., Rajopadhye, S. (2000). Optimizing Memory Usage in the Polyhedral Model. *ACM Transactions on Programming Languages and Systems* 22(5): 773–815.
25. Quinton, P., Robert, Y. (1989). *Systolic Algorithms and Architectures*. Prentice Hall and Masson.
26. Schreiber, R., Aditya, S.G., Rau, B.R., Mahlke, S., Kathail, V., Cronquist, D., Sivaraman, M. (2001). PICO-NPA: High-Level Synthesis of Nonprogrammable Hardware Accelerators. *Journal of VLSI Signal Processing*.
27. Synplify Pro 7.0 Reference Manual, October 2001.
28. Weiss, K., Oetker, C., Katchan, I., Steckstor, T., Rosenstiel, W. (2000). Power Estimation Approach for SRAM-based FPGAs. In: *IEEE Symposium of Field Programmable Gate Array*.
29. Wilde, D. (1993) A Library for doing Polyhedral Operations. Technical Report 785, Irisa, Rennes. France.
30. Wolf, W., Martinez, N. (2002). Session: Platform Based Design and Virtual Component Reuse. In: *Design Automation and Test in Europe (DATE) 2002*. Paris: IEEE Computer Society Press, pp. 296–316.
31. Zhu, X., Lin, B. (1999). Hardware Compilation for FPGA-Based Configurable Computing Machines. In: *Design Automation Conference (DAC)*. New Orleans: ACM, pp. 697–703.

8

Goal-Driven Reconfiguration of Polymorphous Architectures

Sumit Lohani and Shuvra S. Bhattacharyya
University of Maryland at College Park, College Park, Maryland

I. INTRODUCTION

There have been many advancements in recent years on architectures for reconfigurable processing engines (e.g. see [5,10,11]). With the increasing degree of reconfigurability in processing architectures, it is useful to view embedded multiprocessor systems as polymorphous computing architectures (PCAs), in which configurable attributes of the architecture and software are adapted in response to dynamically changing needs. Such attributes may include items such as inter-processor message routing, caching policies, scheduling policies, processor voltages, resource allocation to computing units, and synchronization protocols. A PCA can be a particularly useful platform for developing a computing system where applications and performance requirements change at run-time as one can adaptively configure the PCA to suit the dynamic constraints and objectives.

This chapter takes a step toward bridging techniques for scheduling and system synthesis with reconfigurable processing platforms and the dynamically-changing application requirements that drive these platforms. We first formulate the problem of executing application dataflow graphs on a polymorphous computing architecture such that specified performance requirements are satisfied, where the requirements may vary over time and the application may have tasks with non-deterministic execution times (e.g., due to data dependencies or unpredictable events such as cache misses and interrupts). We analyze key properties of this problem and the complexity of some relevant sub-problems. We then develop a flexible heuristic framework for guiding the run-time configuration adaptation process and show through simulation experiments that this approach can efficiently handle both dynamics in performance requirements and dynamics in task execution time behavior.

In the application model addressed in this chapter, computational tasks (actors), which are represented by dataflow graph vertices, in the application are allowed to have stochastic execution times with static distributions or distributions that may vary slowly over time. The computing unit is a reconfigurable multiprocessor architecture, and the objective is to find a mapping of the actors in the application onto the processors in the multiprocessor and the configuration that the architecture should assume such that performance-related constraints (e.g., constraints on power, resource usage or throughput) are satisfied and objectives (e.g., maximizing throughput or minimizing latency) are optimized effectively. Furthermore, the constraints and objectives may vary over time, and thus, overall solution quality can be viewed in terms of how efficiently reconfiguration of the architecture tracks changes in the applications requirements. Henceforth, we will refer to this problem as the polymorphous computing architecture mapping (PCA mapping) problem. As can be seen, the PCA mapping problem is quite general in nature and even very restricted special cases can be proved to be NP-complete.

The approach suggested in this chapter is correspondingly general and can handle diverse applications and performance requirements. All the reported experiments were performed on an abstraction of the Raw architecture [10] that incorporates salient features of the architecture such as the programmability of interconnects between processors. For experiments, the self-timed execution of applications on this abstracted Raw architecture was simulated using the *interprocessor communication* (IPC) graph model [9].

The emphasis in this chapter is on coordination of the on-line configuration management process for reconfigurable networks of processors, rather than the development of specialized configuration optimization techniques (such as fixed-objective scheduling and allocation), which are already in abundance in the literature (e.g., see [9] for a survey). Our work is complementary to such existing efforts and also to work on multiprocessor system synthesis (e.g., see [1,2]), which can be used to derive the store of precomputed configurations that is input to the techniques developed in this chapter.

II. PROBLEM FORMULATION

A set of relevant metrics, such as latency, throughput, average power, peak power, and number of resources, is denoted by M. If a certain metric appears as a constraint with a value to be satisfied when the application executes, then this metric is referred to as a constraint metric and the value as a constraint value for that particular metric. A constraint value belongs to the set of real numbers. A pair of constraint metrics and constraint values is called a constraint pair. A sequence of constraint pairs, in turn, is referred to as a constraint vector, and is denoted by

$$V = [(m_1, c_1), (m_1, c_2), \ldots, (m_K, c_K)] \tag{1}$$

where $m_1, m_2, \ldots, m_K$ represent any metrics in M, and $c_1, c_2, \ldots, c_K$ represent the corresponding constraint values, for $K \in \{0, N\}$, where N is the number of all constraint pairs. This (possibly empty) sequence of constraint pairs in a constraint vector is prioritized such that (m_i, c_i) is a higher priority constraint pair than a constraint pair (m_j, c_j) if $I < j$, for $i, j \in \{1,2,\ldots,K\}$ in a constraint vector $V = [(m_1, c_1), (m_1, c_2), \ldots, (m_K, c_K)]$. A metric m_R that is to be optimized after all constraints have been satisfied is called a *residual objective*. A *goal* g is an ordered pair (V, m_R), where V is a constraint vector and m_R is a residual objective. If there is no residual objective, then the goal is composed of only a constraint vector and can be represented by $(V, \perp)$. Here, the symbol $\perp$ represents the absence of a residual objective. Also, without loss of generality, the metrics are such that the associated optimization problems are to *minimize* the respective metric (i.e., a lower value of a metric is always better than a higher value). Metrics for which higher values are more desirable must thus be transformed into corresponding metrics for which lower values are better. For example, in iterative applications, the throughput (average rate of completion of application iterations) can be recast as the *average iteration period*, which is the reciprocal of the throughput.

Example 1. Consider a set of relevant metrics $M = \{L,P,T\}$, where L is the latency, P is the average power consumption, and T is the average iteration period. Consider the goal

$$g = [(L, 50), (P, 100), (L, 40), (P, 70), T]$$

In g, the constraint pair $(L,50)$ has higher priority than the constraint pair $(P,100)$, which in turn has higher priority than the constraint pair $(L,40)$. The metric T is the residual objective.

This definition of reconfiguration goals as prioritized lists with optional residual objectives leads to a view of dynamic reconfiguration as a sequence of one-dimensional optimization problems. This simplification is useful because run-time adaptation techniques must be of relatively low complexity, and thus, one-dimensional optimization is a better match. Additionally, it allows us to leverage existing libraries of single-dimensional synthesis techniques, which are more abundant than multidimensional techniques.

Third, it provides an intuitive and unambiguous format for designers to prioritize multidimensional application requirements. Note, however, that this formulation applies only to run-time reconfiguration, and multi-dimensional optimization techniques, such as SPEA-based methods [12], can be used off-line in arbitrary ways to compute caches of pre-computed configurations. Use of such caches will be discussed further in Sections III–V.

For example, in Example 1, we initially have an unconstrained latency optimization problem (since the first constraint involves latency). As we adapt the system configuration with techniques that address this problem, we will in general improve the latency. Once the latency improves to 50 time units, the current constraint is satisfied, and we switch to a power-optimization problem subject to a constraint of $L = 50$. The optimization process may continue in this manner until the last constraint is satisfied (in this case, $P = 70$), at which point run-time adaptation stops (if there is no residual objective) or reaches a terminal mode of optimizing the residual objective subject to all constraints in the constraint vector. This mode then continues until the system shuts down or the application goal changes.

Mapping an application to a multiprocessor architecture includes defining a task-to-processor mapping along with defining the configuration of the reconfigurable architecture. In this chapter, the scope of the word configuration is expanded to also include the mapping of the application onto the reconfigurable architecture. Therefore, a *configuration* consists of two components, (1) task-to-processor mapping, and (2) configuration of the architecture. Thus, the word configuration is used in the above sense, unless stated otherwise. A given application, goal, and resource set define an instance of the PCA mapping problem. Input to the model is an *instance* that may change with time. We define the *design space* as the set of all feasible combinations of an instance and a configuration. The *solution space* for a feasible instance is the set of all feasible configurations for that instance. Latency, average iteration period, average power, and peak power are some of the commonly encountered metrics. With many metrics of simultaneous relevance, the goal space is too vast to be fully explored before run-time, and run-time adaptation of configurations is generally advantageous.

Figure 1 illustrates a general model for solving the PCA mapping algorithm with a combination of off-line and on-line techniques. The main components of the model are the *off-line component*, the *configuration store (CS)*, and the *on-line component*. The off-line component, whose objective is to pre-compute a set of efficient candidate mappings for various run-time scenarios, can be constructed using existing methods for scheduling, system synthesis, and multi-objective optimization. The focus of this chapter is thus on the on-line refinement component and its interaction with the configuration store.

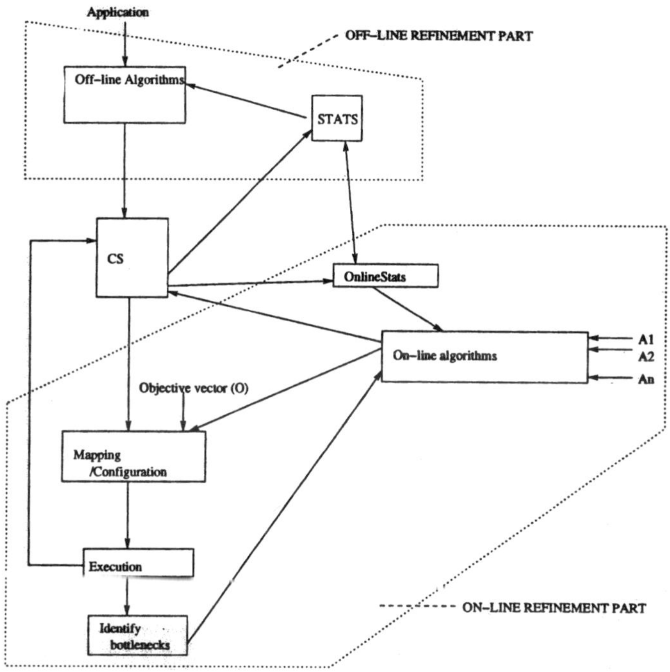

Figure 1 An overview of the system-level reconfiguration framework studied in this chapter.

For a given instance, not every configuration is suitable as some configurations may violate constraints or may not adequately address residual objectives. As the goal changes for a given application, the system needs to derive a suitable adaptation of the run-time configuration. Optimally solving this problem is undecidable in many contexts. Also, reconfigurability of the architecture and the stochastic variance of execution times greatly complicates the solution space consisting of all possible configurations for the input of a goal and a given application. Since computing a suitable configuration is performed during the execution of an application, one can not apply exhaustive or relatively sophisticated search strategies as those techniques will take away excessive computational resources away from the application itself. To address this trade-off (thoroughness of dynamic optimization vs. resources drained from the application), our model of the PCA mapping problem also accounts for the time spent in computing efficient adaptations of mappings at run-time on the basis of feedback obtained from execution and identification of bottlenecks, and thus always tries to move toward an optimal solution. This is taken care of in the on-line refinement part of the model, which consists of low-complexity algorithms that find and refine configurations for a given instance. It also consists of feedback units shown by the "*Identify bottlenecks*" block in Fig. 1 that takes feedback from the execution of the configurations and modifies the configurations so as to better suit the active goal. The *OnlineStats* unit in the on-line refinement part of the model stores short-term statistics that can be used by on-line algorithms.

A configuration store is used to store high-quality points in the design space that have been explored so that one can use them later as need be. Thus, a configuration store keeps tuples of the form {`App`,g,c}, where `App` represents an application, g represents a goal for that application, and c represents a configuration that satisfies the goal g. Henceforth, storing a configuration in the configuration store and storing a goal in the configuration store have been used interchangeably to mean storing a tuple of the above stated form in the configuration store, unless stated otherwise. The off-line refinement part of Fig. 1 consists of high-complexity algorithms that yield better solutions. It is acceptable for them to be of high-complexity as they are used off-line, and do not compete for resources with the application. In Fig. 1, the *STATS* unit stores statistics about the application (e.g., distributions of execution times for different actors frequencies of occurrence of some particular regions of the goal space, etc.). Off-line algorithms use these statistics to explore the solution space for input instances. As soon as the goal or application changes, an initial configuration is found using the on-line configuration management component in conjunction with the configuration store. On-line algorithms keep improving the configuration that is being executed, using the feedback from the execution. In the meantime, off-line algorithms may keep exploring areas of the design space and merge the relevant information into the configuration store (for use in the selection of future initial configurations).

III. CONFIGURATION MANAGEMENT MODEL

The overview of our PCA system synthesis model shows that it is very adaptive in nature and thus is suitable for applications with stochastic execution times and time-varying goals. This section develops further details of this model.

A. Evaluation of Configurations and Goals

It is useful to define some measure of how well a given configuration executes for a particular instance. This evaluation measure should allow unambiguous comparison between two configurations based on the current goal.

Suppose we are given a goal $g = [V, m_R]$, where

$$V = [(m_1, c_1), (m_1, c_2), \ldots, (m_K, c_K)] \tag{2}$$

We define the *quality* of a system configuration C with respect to goal g, denoted $Q_g(C)$ (or simply $Q(C)$ if g is understood) as the ordered pair $Q(C) = (k,v)$, where $k + 1$ is the index of first unsatisfied constraint in the constraint vector of g (i.e., the lowest-index constraint in g that is not satisfied by the configuration), and v is the value obtained for the metric m_{k+1}. If configuration C satisfies all n constraints in the constraint vector of g, then we say that C *satisfies* g, and in this case, $Q(C) = (n + 1, v_R)$, where v_R is the value obtained for the residual objective m_R if $m_R \neq \perp$ or $v_R = -\infty$ if $m_R = \perp$.

In summary, the quality of a configuration measures a configuration with respect to a given goal, and given a goal and two configuration C_1 and C_2 with qualities $Q(C_1) = (k_1,v_1)$ and $Q(C_2) = (k_2,v_2)$ for that instance, respectively, C_1 has higher quality than C_2 if

$$(k_1 > k_2) \ or \ ((k_1 = k_2) \ and \ (v_1 < v_2)). \tag{3}$$

B. Configuration Store

A configuration store serves as a repository of alternative configurations. A configuration store can be divided into several sub-stores (sub-CSs), one for each relevant application. Each sub-CS has some configurations stored in it, one for a specific combination of goal and resource set. In the later part of this section, we assume that we are dealing with a fixed application and a fixed resource set, unless stated otherwise. This does not detract from the applicability of the ideas developed later as they can be generalized to include various applications and resource sets using the hierarchical model of configuration store explained above.

Assuming a fixed application and resource set, selecting the goals whose corresponding configurations should be stored in the configuration store depends on various factors such as the size of the configuration store, the optimality of the stored configuration, computational resources drained from the application during execution by the on-line refinement algorithms, and the expected or observed frequency of specific goals.

C. Acceptability of Configurations

Notions of *acceptability* and *cover* emerge naturally from this concept of configurations stores, and guide the construction and adaptation of the configuration store in our model. For example, one can envision the reconfiguration process as selecting an acceptable configuration and gradually tightening the notion of acceptability to guide the on-line refinement process. The following definition makes these notions precise.

Definition 1 Given two goals g_1 and g_2, we say that g_1 is *acceptable* for g_2, denoted $g_1 \rightarrow g_2$, if a configuration that satisfies g_1 is an acceptable implementation for g_2. If $g_1 \rightarrow g_2$, we also say that g_1 *covers* g_2. Given a set Φ of goals and a specific goal g, the space of g over Φ (or simply, the *space* of g, if Φ is understood) is $\{g' \varepsilon \Phi | g \rightarrow g'\}$. Thus, the space of a goal g is the set of goals that are acceptably implemented by any configuration that satisfies g. The space of a goal g is represented by space(g).

The following result, proved and elaborated in [7], shows that the acceptability of configurations is a particularly well-behaved relation if it is a partial order.

Theorem 1 If we have a finite set Φ of relevant goals, and the acceptability relation is a partial order, then there exists a unique, minimal set of goals $\{g_1, g_2, \ldots, g_n\}$ such that

$$\bigcup_{i=1}^{n} space(g_i) = \Phi \tag{4}$$

and this set of goals can be computed in polynomial time in $|\Phi|$, the number of relevant goals.

Definition 2 Dominance relation: A point $p \in \Re^n$ dominates a point $q \in \Re^n$ if $p_i \leq q_i \forall i = 1, \ldots, n$, where p_i and q_i denote the ith components of p and q, respectively.

One can see that the dominance relation is a partial order [4]. We can have an acceptability relation between goals based on the dominance relation where a goal g_1 is acceptable for a goal g_2 if the constraint vector of the goal g_1 dominates the constraint vector of the goal g_2, and the residual objectives for both goals are same. The following example illustrates an acceptability relation that is not a partial order.

Example 2 Suppose that we have a single constraint metric, which is the average iteration period T of the system. Thus, the constraint associated with a goal g can be expressed as the desired average iteration period $T(g)$. Suppose that in a particular implementation context, the acceptability relation $g_1 \rightarrow g_2$ is defined by $T(g_1) - T(g_2) \leq \Delta T$ for some positive real number ΔT. Thus, a configuration for g_1 can be worse than what is desired under g_2, and still be acceptable for g_2, as long as the deviation does not exceed the threshold ΔT. Suppose also that the goals g_1, g_2, and g_3 have desired average iteration period values of

$$T(g_1) = 5, T(g_2) = 5 - \frac{3\Delta T}{4} \text{ and } T(g_3) = 5 - \frac{3\Delta T}{2} \tag{5}$$

One can then see that $g_1 \rightarrow g_2$ and $g_2 \rightarrow g_3$ but g_1 is not acceptable for g_3. Therefore, this acceptability relation is not transitive and thus is not a partial order.

An acceptability relation between goals based on the dominance relation or any other partial order leads to valuable properties such as that exposed by Theorem 1. Also, the dominance relation is a natural candidate for an acceptability relation among goals, as a configuration corresponding to the dominating goal can be used in place of a configuration corresponding to the dominated goal without violating any constraints. This motivates our use of the dominance relation in managing configuration stores. One can observe that our approaches of defining a goal and the quality of a configuration are all consistent with acceptability based on the dominance relation.

IV. ON-LINE CONFIGURATION MANAGEMENT

In this section, we define an on-line configuration management framework called *CMF* that defines how to choose an initial configuration for a particular instance, and how the on-line adaptation for that configuration should proceed. We also formulate problems related to storage of configurations in the configuration store. These problems and our models to solve them provide fundamental analysis of the complexity of configuration management and provide feasible, low-complexity solutions to this problem.

A pseudocode outline of the CMF approach is shown in Fig. 2. The objective is to provide a framework that imposes minimal constraints on how reconfiguration is actually performed, while providing systematic support for managing the reconfiguration process in terms of configuration stores, performance constraints, and optimization objectives. CMF is a meta-algorithm because specific details of the architecture, the application, and the on-line adaptation algorithms are left unspecified, and can be customized based on the relevant classes of applications and architectures. This meta-algorithm maintains a *current objective* at all times, where the goal is always to improve the current objective without violating any of the previously satisfied constraints. The function *onLineAdaptation* takes an objective metric, a constraint value, and a configuration as inputs, and keeps

```
function CMF
/* Global variables accessed: the current goal and the
set of pre-computed configurations, respectively. */
global goal g_c = [(m_1, c_1), (m_2, c_2), ..., (m_K, c_K), m_R],
global configurationStore C, global stack S=emptyStack,
global goal g, g_0, global constant time timelimit

g = g_0 = g_c
/* The current optimization metric and the current
constraint to satisfy */
objective objective_c = m_R
constraint constraint_c = null
σ = {c∊C|c satisfies g}
while (σ == φ) {
    (g, constraint_c, objective_c) = demoteConstraint(g, S)
    σ = {c∊C|c staisfies g}
}
/* Select an admissible configuration from σ using
some heuristic or specialized, optimal algorithm. */
configuration_c = select(σ)
/* Keep trying to refine the current configuration
according to the current goal until the goal changes
(g_c may change at any time under external control). */
while (g_c == g_0) {
    onLineAdaptation(objective_c, constraint_c, configuration_c)
    adjustAdaptation
}
end function
```

Figure 2 The CMF framework for goal-driven reconfiguration. Details of the *adjustAdaptation* phase are elaborated in Fig. 3.

refining the configuration in an effort to continually improve its quality (as defined in Section III). This function would typically be called within an enclosing loop that performs any system-dependent re-initialization and re-invokes the function immediately after the previous invocation of the function terminates (observe that the function terminates when the current goal is changed).

```
if ((constraint_c is satisfied) &&
        (objective_c is not the residual objective)) {
   /* Move to the next unsatisfied constraint or to
   the residual objective */
   (g, constraint_c, objective_c) = promoteConstraint(g, S)
} else if (objective_c is the residual objective) {
   continue
} else if (constraint_c is not satisfied due to
            timelimit restriction) {
    warning(timelimit insufficient to satisfy
                constraint)
    /* timelimit handler code */
}
```

Figure 3 Sketch of the *adjustAdaptation* block for the CMF framework of Fig. 2.

Pseudocode for the related functions is given in Fig. 4. We have implemented CMF, and simulation results pertaining to it are discussed in Section V.

A. Issues Related to Configuration Management

Before proceeding with discussion of our experiments with CMF, we first study some fundamental versions of the problems related to configuration management, discuss their complexity, and relate aspects of them to well-studied problems. Two related problems regarding the size of the configuration store are as follows.

P1. Find the minimum size configuration store and the goals whose configurations should be stored in it such that all relevant goals are covered.

P2. If one has a well-defined measure of "distance" between goals and the goal space is a metric space [3], then for a given fixed size configuration store, find the goals whose configurations should be stored such that the sum of the distances of those goals that are not present in the configuration store, from the distance-wise nearest goal present in the configuration store, is minimum.

The following definitions are be helpful to analyze the complexity of problem P1 and P2.

Definition 3 For a directed graph $G(V,E)$, a subset D of V is a *dominating set* if $\forall v \in V$, either $v \in D$ or there exists $u \in D$, such that $(u, v) \in E$.

```
function promoteConstraint
input goal g = [(m_1, c_1), (m_2, c_2), ..., (m_{K-1}, c_{K-1}), m_K]
output goal, constraint, objective
goal g'
constraint v = S.pop()
objective m = S.pop()
if (S is not empty){
    constraint x = S.pop()
    S.push(x)
} else {
    constraint x = -∞
}
g' = [(m_1, c_1), (m_2, c_2), ..., (m_{K-1}, c_{K-1}), (m_K, v), m]
return (g', x, m)
end function

function demoteConstraint
input goal g = [(m_1, c_1), (m_2, c_2), ..., (m_K, c_K), m_R]
output goal, constraint, objective
goal g'
S.push(m_R)
S.push(c_K)
g' = [(m_1, c_1), (m_2, c_2), ..., (m_{K-1}, c_{K-1}), m_K]
return (g', c_K, m_K)
end function
```

Figure 4 Definition of functions `promoteConstraint` and `demoteConstraint` from Fig. 2.

Definition 4 Minimum dominating set problem. Given a directed graph, find a minimum dominating set of the graph.

Definition 5 k-median problem. In the k-median problem, we are given a set of potential facility locations F. Any open facility can provide an unlimited amount of a certain commodity. There is a set of clients or demand points D that require service; client $j \in D$ has a positive demand of commodity d_j that must be shipped from one of the open facilities. If a facility at location $i \in F$ is used to satisfy the demand of client $j \in D$, the service or transportation cost incurred is proportional to the distance c_{ij} from i to j. This distance function c is non-negative, symmetric, and satisfies the triangle inequality. The goal is to determine k potential facility locations at which to open facilities and an assignment of clients to these facilities so as to minimize the overall service cost.

Definition 6 Facility location problem. In the facility location problem, we are given a set of potential facility locations F; building a facility at location $i \in F$ has an associated non-negative fixed cost f_i; and any open facility can provide an unlimited amount of a certain commodity. There is a set of clients or demand points D that require service; client $j \in D$ has a positive demand of commodity d_j that must be shipped from one of the open facilities. If a facility at location $i \in F$ is used to satisfy the demand of client $j \in D$, the service or transportation cost incurred is proportional to the distance c_{ij} from i to j. The distance function c is non-negative, symmetric, and satisfies the triangle inequality. The goal is to determine a subset of the set of potential facility locations at which to open facilities and an assignment of clients to these facilities so as to minimize the overall total cost.

P1 and P2 can be viewed, respectively, in terms of the well-known problems of minimum dominating sets and k-medians. To reduce P1 from the minimum dominating set problem [4], for every vertex in the dominating set problem, instantiate a goal, and for every edge, instantiate a condition that the goal corresponding to the source vertex is acceptable to the goal corresponding to the sink vertex. The problem P1 related to this set of goals and the acceptability relation among goals is equivalent to the given minimum dominating set problem instance. The vertices in the given minimum dominating set problem instance, corresponding to the goals that should be stored in the configuration store (found by solving P1) constitute a minimum dominating set for the given minimum dominating set problem instance. This can be used to show that the problem P1 is NP-hard (see [7] for more details). However, if the acceptability relation is a partial order, then the minimum dominating set can be found in polynomial time by picking up all the vertices with no incoming edges in the graph of the minimum dominating set problem. This is in accordance with Theorem 1 and further underscores the advantage of using acceptability relations that are partial orders.

If the associated distance function is defined between any two goals and the goal space is a metric space, then problem P2 can be modeled in terms of the *k-median problem* [2,6], as shown in [7]. For the simple case of a two-dimensional goal space, a polynomial-time approximation algorithm with a 3-approximation factor exists for the k-median problem [2].

Configuration management problems P1 and P2 can be viewed as extreme cases in the sense that in one of them we want to cover all feasible goals without considering how large the minimum size configuration store would be (P1), and in the other case, we have a fixed-size configuration store and we are trying to find out the maximum number of goals that can be covered using that configuration store even though that number could be much less than the total number of relevant goals (P2). A more elaborate formulation would be one in which we have to pay extra cost for increasing the size of the configuration store, but in doing so we would gain some additional service by being able to store more goals in the configuration store. This way we can explore various trade-offs between the size of the configuration store versus the number of goals stored in a well-defined way. For the specific case when a distance function is defined between any two goals and the goal space is a metric space, these trade-offs can be explored by modeling this problem as a facility-location problem [3,6,8], as explained in [7]. A polynomial time algorithm with an approximation guarantee of 1.74 exists for the facility location problem [3].

V. ON-LINE ADAPTATION

In this section, we focus on the metrics of average iteration period and power consumption, and develop low-complexity, on-line strategies based on heuristics for average iteration period optimization and power optimization as implementations of the function *onLineAdaptation* in Fig. 2. Note that the average iteration period optimization is analyzed to throughput optimization as average iteration period is defined as the reciprocal of throughput. The objective is to demonstrate the efficacy of the CMF model and show that it can produce efficient tracking of time-varying application requirements.

The approach of taking feedback from the execution of the application makes these on-line methods able to handle even applications with stochastic execution times that have time-varying distributions, in addition to applications with fixed execution times and applications with stochastic attributes that have stationary distributions. In general, this on-line refinement formulation can thus be viewed as an approach to tracking the dynamics of the goal and the characteristics of the application.

To experiment with CMF, we used a simple heuristic based on load balancing [13] to optimize average iteration period during online adaptation. Pseudocode for this heuristic is represented by function *adaptT* in Fig. 5. In the pseudocode, *moveTask*(*c*,*n*) is a function that chooses *n* tasks from a maximally loaded processor in a configuration *c* and, randomly, moves them to appropriate locations on a minimally loaded processor, and returns the modified configuration. Randomization in choosing tasks from the maximally loaded processor provides a low-complexity approach to increase the explored region of the design space and to calibrate the configuration to dynamic application characteristics. The function *executeT*(*c*,*l*) is a function that executes the application according to configuration *c* for a time interval

```
/* This function adapts the given input configuration
while executing the application. */

function adaptT
input configuration c, constraint v
global constant time l

Average iteration period T_old = executeT(c, l)
time t
configuration c_old = c
n = 1

while ((clock < timelimit) && (constraint valu  v is not
        satisfied)) {
    if ( Bookkeeping related to moveTask shows that all
    n-task movements on c_old have been tried on c_old
    using function moveTask without any improvement) {
        n = n + 1
    }
    c = moveTask(c_old, n)
    T = executeT(c, l)
    clock=clock + l
    if (T ⩽ T_old) {
        c_old = c
        T_old = T
        n = 1
    }
}
end function
```

Figure 5 An online adaptation approach for throughput optimization.

of length l, and returns the average iteration period of the application during that interval. The value of l depends on the non-determinacy of the application. We define the non-determinacy of an application in the following way.

Let the number of possible execution times taken by an actor i be denoted by n_i. We denote the set of n_i possible execution times taken by an actor i as $\{t_{i1}, t_{i2}, \ldots, t_{in_i}\}$. The probability of occurrence of a possible execution time t_{ik} for actor i, is denoted by p_{ik}, for all $k = 1, \ldots, n_i$. The *degree of non-determinacy* λ is a measure of the overall amount of non-determinacy in the application, specifically, in the actor execution times, and is defined as

$$\lambda = \frac{\sum_i \left\{ \sum_{k=1}^{n_i} \{p_{ik}(t_{ik} - t_{i,mean})^2\} \right\}}{\sum_i \{t_{i,mean}\}^2} \tag{6}$$

where $t_{i,mean}$ denotes the mean execution time of actor i, and is defined as

$$t_{i,mean} = \left(\sum_{k=1}^{n_i} t_{ik} \right) / n_i \tag{7}$$

Generally, the more non-deterministic the application is, the longer it needs to be executed to determine an accurate value of average iteration period.

The function *adaptT* returns a configuration that it deems most appropriate for average iteration period maximization. Note that if moving any single task from the maximally loaded processor to the minimally loaded processor does not improve performance then the heuristic chooses a *pair* of tasks to be moved to another processor. This approach of progressively increasing the number of tasks to be moved continues whenever all combinations for a particular number of tasks have been exhausted. Thus this approach attempts to make small low-complexity changes first and if that does not improve performance, the approach gradually reaches toward higher-complexity changes. The higher complexity changes are larger in number than small, low-complexity changes, and help the system in escaping from local minima.

In our experiments, inter-processor communication (IPC) per time unit during the execution is taken as an estimate for relative power consumption. Since IPC consumes relatively large amounts of power, it is a reasonable approximation for comparing the power consumption levels of alternative configurations on a homogeneous multiprocessor. To find a configuration that reduces the power consumption, we use an approach (called *adaptPower*) similar to the *adaptT* approach used for average iteration period optimization, except that the probability of a task on a maximally loaded processor being transferred to a minimally loaded processor depends upon the IPC associated with that task. The higher the IPC associated with a task, the higher its chances are of being transferred to another processor.

VI. EXPERIMENTAL RESULTS

An on-line adaptation scheme for refining a given goal is specified in Fig. 6, and it is represented as function *onlineAdaptation* in the CMF pseudocode

```
function onLineAdaptation
input objective_c, constraint_c, configuration_c

time t = 0
while ((t < timelimit and g_c = g_0) && (constraint_c is not
        satisfied)) {
    if (objective_c = average iteration period)
      adaptT (configuration_c, constraint_c)
    else if (objective_c == power)
      adaptPower(configuration_c, constraint_c)
    else if ...
      ... /* adapt for other objectives */ ...
}
end function
```

Figure 6 On-line adaptation scheme. This is an elaboration of function *onLine Adaptation*, which is described in Fig. 2. It is effectively a wrapper for specialized reconfiguration optimizations.

of Fig. 2. In Fig. 6, the appropriate online optimization strategy, such as the *adaptT* or *adaptPower* approaches discussed above, is selected depending on the current optimization objective and system state. Typically, this strategy will be drawn dynamically from a library of simple, low-complexity techniques.

Table 1 shows the performance of our implementation of CMF using the heuristics developed in Section V for average iteration period optimization and power optimization based on various goals applied to several DSP benchmarks, including fast Fourier transform, filter bank, music synthesis, and measurement applications. The starting configuration that is refined is found by using standard critical path scheduling. The critical path length is computed in terms of average execution times of actors. The set of relevant metrics M for our experiments is $M = \{T,P\}$, where T denotes the average iteration period of the execution and P denotes the average power consumption. Experiments are reported for the following eight goals.

$$g_1 = \{(P, 0.270), (T, 265), (P, 0.250), (T, 255), P\}$$
$$g_2 = \{(T, 260), (P, 0.240), T\}$$
$$g_3 = \{(P, 0.125), (T, 180), P\}$$
$$g_4 = \{(T, 165), (P, 0.110), (T, 160), P\}$$
$$g_5 = \{(T, 360), (P, 0.160), (T, 355), (P, 0.155), (T, 350), P\}$$

Table 1 Experimental Results for CMF

Application	λ	Goal	Metric	v_0	v_{10}	v_{20}	v_{30}	v_{40}	v_{50}	v_{60}
fft1	0	g_1	T	278	278	278	278	256	254	254
			P	.273	.269	.269	.269	.204	.226	.226
fft1	.359	g_2	T	309	256	251	251	251	252	259
			P	.242	.282	.278	.278	.278	.257	.221
qmf	0	g_3	T	145	242	198	198	186	170	170
			P	.133	.117	.098	.098	.088	.096	.096
qmf	.256	g_4	T	142	164	162	162	153	153	153
			P	.136	.127	.110	.110	.110	.110	.1100
karp	0	g_5	T	395	353	346	342	342	342	342
			P	.131	.158	.156	.148	.148	.148	.148
karp	.309	g_6	T	450	352	300	342	342	346	346
			P	.115	.155	.159	.151	.151	.148	.148
meas	0	g_7	T	220	212	201	184	184	184	184
			P	.054	.075	.059	.021	.021	.021	.021
meas	.405	g_8	T	185	218	212	212	212	210	196
			P	.064	.018	.037	.037	.037	.019	.040

$$g_6 = \{(T, 345), P\}$$
$$g_7 = \{(T, 215), (P, 0.040), T\}$$
$$g_8 = \{(P, 0.053), (T, 215), (P, 0.050), (T, 210), P\}$$

In Table 1, the column entitled "Goal" represents the goal that is applied to the application. Also, for a non-negative integer k, column v_k denotes the value of a metric of the best configuration found by the on-line adaptation scheme, after k configurations have been assessed by executing them for some time. For the same experiments that are reported in Table 1, Table 2 shows the times at which different constraints associated with the applied goals are satisfied. For a given goal that is applied to an application,

Table 2 Results for CMF Tracking an Applied Goal

Application	λ	Goal	n_1	n_2	n_3	n_4	n_5
fft1	0	g_1	1	37	39	42	—
fft1	.359	g_2	7	56	—	—	—
qmf	0	g_3	8	48	—	—	—
qmf	.256	g_4	0	13	36	—	—
karp	0	g_5	4	7	9	28	28
karp	.309	g_6	16	—	—	—	—
meas	0	g_7	8	28	—	—	—
meas	.405	g_8	3	17	17	48	—

n_i denotes the number of configurations that were executed in order to assess them before the ith constraint in the applied goal was satisfied. One can see that in these experiments, CMF is able to meet the constraints specified in the given goals within a reasonable number of configurations.

VII. CONCLUSION

In this chapter, we have developed a framework called CMF for on-line adaptation of system wide configurations of embedded multiprocessors. The objective was to provide a framework that imposed minimal constraints on how reconfiguration is actually performed (i.e., the specific optimization algorithms that are used during offline and online configuration synthesis), while providing systematic support for managing the reconfiguration process in terms of configuration stores, performance constraints, and optimization objectives. The CMF approach was shown to be effective through analysis and experimental results on several DSP benchmarks, which demonstrated the ability of CMF to systematically adapt system configurations toward progressively better solutions for a variety of goals, even in the presence of significant uncertainties in task execution times.

VIII. ACKNOWLEDGMENTS

This research was sponsored in part by the U.S. National Science Foundation (9734275) and by DARPA (contract #F30602-01-C-0171, through the USC Information Sciences Institute).

REFERENCES

1. Bhattacharyya, S. S. (2002). Hardware/software co-synthesis of DSP systems. In: Hu, Y.H., ed. *Programmable Digital Signal Processors: Architecture, Programming, and Applications*. Marcel Dekker, Inc., pp. 333–378.
2. Blickle, T., Teich, J., Thiele, L. (1998). System-level synthesis using evolutionary algorithms. *Journal of Design Automation for Embedded Systems* 3(1):23–58.
3. Chudak, F. (1998). Improved approximation algorithms for uncapaciateted facility location. In: Boyd, E. A., Bixby, R. E., Rios-Mercado, R. Z., eds. *Integer Programming and Combinatorial Optimization*. Springer LNCS, pp. 180–194.
4. Cormen, T. H., Stein, C., Leiserson, C. E., Rivest, R. L. (2001). *Introduction to Algorithms*. 2nd ed. MIT Press.
5. Hauser, J. R., Wawrzynek, J. (1997). Garp: A MIPS processor with a recon-

figurable coprocessor. In: *Proceedings of the IEEE Symposium on FPGAs for Custom Computing Machines*, pp. 24–33, April.

6. Jain, K., Vazirani, V. V. (1999). Approximation algorithms for metric facility location and k-median problems using the primaldual scheme and Lagrangian relaxation. In: *Proc. Foundations of Computer Science.*
7. Lohani S., Bhattacharyya S. S. (2002). System synthesis for polymorphous computing architectures. Technical Report UMIACSTR-2002-12, Institute for Advanced Computer Studies, University of Maryland at College Park. Also Computer Science Technical Report CS-TR-4330, February.
8. Shmoys, D. B., Tardos, E., Aardal, K. I. (1997). Approximation algorithms for facility location problems. In: *Proc. 29th ACM Symp. on Theory of Computing*, pp. 265–274.
9. Sriram, S., Bhattacharyya, S. S. (2000). *Embedded Multiprocessors: Scheduling and Synchronization.* Marcel Dekker, Inc.
10. Waingold, E., et al (1997). Baring it all to software: RAW machines. *IEEE Computer Magazine*, pp. 86–93, September.
11. Wong, S., Vassiliadis, S., Cotofana, S. (2001). Microcoded reconfigurable embedded processors: Current developments. In: *Proceedings of the International Workshop on Systems, Architectures, Modeling, and Simulation*, pp. 207–223, July.
12. Zitzler, E., Thiele, L. (November 1999). Multiobjective evolutionary algorithms: A comparative case study and the strength Pareto approach. *IEEE Transactions on Evolutionary Computation* 3(4):257–271.
13. Zomaya, A. Y. (1996). Parallel and distributed computing: The scene, the props, the players. In: Zomaya, A. Y., ed. *Parallel and Distributed Computing Handbook.* New York: McGraw-Hill, pp. 5–23.

9

Realizations of the Extended Linearization Model

Alexandru Turjan, Bart Kienhuis, and Ed F. Deprettere
Leiden Embedded Research Center, Leiden University, Leiden, The Netherlands

I. INTRODUCTION

An appealing and fruitful methodology to deal with exploration or designing *applications–architectures pairs* has become known as the *Y-chart approach*, [1]. This approach embraces two fundamental notions: the *separation of concerns* and the *abstraction hierarchy*. The concerns are: the application, the architecture, and the mapping. The abstraction hierarchy, introduced in [2] as the *abstraction pyramid*, bridges—whether for exploration or synthesis purposes—the gap between high-level application specification and low-level architecture specification by defining a number of abstraction levels and a corresponding *stack of Y-charts*. At each level, application models, architecture models, and mapping models must match to make exploration and synthesis feasible.

Several research groups around the globe are currently experimenting with this methodology, some explicitly and others implicitly. They are, naturally, all focusing on different application domains that lead to different views on this methodology. Applications in the realm of automotive, multimedia, and communications have different requirements, constraints, and boundary conditions which result in different challenges.

The Leiden Embedded Research Group focuses on applications that can be specified as parameterized affine nested loop programs (NLPs). The group has been developing and implementing the Compaan tool chain to translate such applications from their imperative language specification into

Kahn process networks (KPN) [3]. The application specification language is Matlab or C, and the tool-chain is a compiler through which a range of KPNs can be obtained for any given application specified as a parameterized NLP.

The processes in the Compaan-generated KPNs are not (completely) specified in an imperative model of computation because the distance between that model and the models in which architecture components—in particular the processing units—are specified is too large. This is not specific to the application domain for which Compaan is an appropriate translation tool set; it is a problem that is revealed wherever the Y-chart methodology is used. Of course, the processes in KPNs may be specified in terms of more than one model of computation. For example, one could be obtained for the *control data flow graphs* model [4] or for one or more dataflow network models [5].

The Process Network Model (PN) in Compaan is the Kahn process network (KPN) model [6], which consists of concurrent autonomous processes that communicate in a point to point fashion over unbounded FIFO channels using a blocking–read synchonization. The strength of a process network is that it uses no global memory and no global scheduler. This makes KPN very appealing for further implementation into hardware [7].

In the Compaan KPN processes, each process executes an internal function following a local schedule. At each execution (also referred to as *iteration*) this function reads/writes data from/to different FIFOs. An input port domain (IPD) of a process is the union of the iterations at which the process's function reads data from the same FIFO. An output port domain (OPD) of a process is the union of the iterations at which the function writes data to the same FIFO. Each FIFO uniquely relates an input port to an output port forming to an instance of the classical *producer/consumer* pair [8].

One of the tools in the Compaan tool chain is *Panda*. Panda accepts as input the description of a polyhedron reduce dependence graph (PRDG) and transforms this PRDG into a process network. This transformation is done in a number of steps. One of the steps involved is *linearization*, in which a high dimensional data structure (e.g., matrix $A[i,j]$) is linearized into a single linear stream of data. In case of Kahn process networks, the linearization model is a FIFO buffer as shown in Fig. 1. In the top part of this figure, a producer and consumer process communicate the data array A [i,j] using global memory. In the linearization step, this communication is replaced with a FIFO buffer, leading to the producer/consumer processes given in the lower part of Fig. 1. Observe that in the top part, the indices i and j are used to address matrix A. In the bottom part, the reference to A has been eliminated. The for-loops only describe an order and data produced by the function and is placed on the FIFO buffer. There are cases, however, in which a FIFO as the linearization

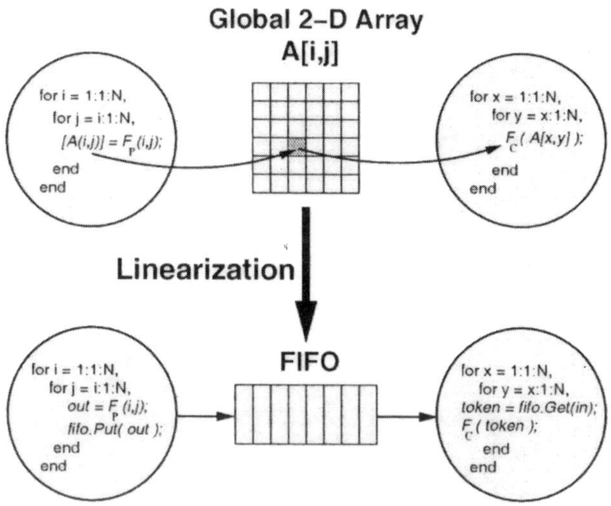

Figure 1 The standard linearization model.

model (LM) no longer holds. If the order data produced is different from the order data needed to be consumed, a FIFO buffer no longer is enough. In [9], we proposed an extension to the LM, which we called the *extended linearization model* (ELM). This model includes an additional reordering mechanism that consists of a *controller* unit *and* some *reordering memory*. The ELM preserves the semantics of the KPN model. As we will show in this chapter, the ELM can be realized in different ways and each realization has its own strength and weakness. Based on these realizations, alternative hardware/software mappings of the Compaan-generated network onto different platforms are feasible.

II. IN ORDER/OUT OF ORDER CASE

Consider the two KPN processes in Fig. 2 with node domains P = {p, I} and C = {c, K}, respectively that are collections of atomic nodes p(i, j) and c(x,y) defined on the domains $I = \{(i,j)|\ 3 \le j \le N \wedge 1 \le i \le N-2\}$ and $K = \{(x,y)|\ 2 \le x \le N-1 \wedge 2 \le y \le N-1\}$, respectively. In the first process one of the OPDs is O = {out, J} that is a collection of atomic output ports out(i,j) defined on the domain $J = \{(i,j)|\ 3 \le j \le N \wedge i+2 \le j \wedge 1 \le i \le N-2\}$. In the second process one of the IPDs is I = {in, L} that is a collection of atomic input

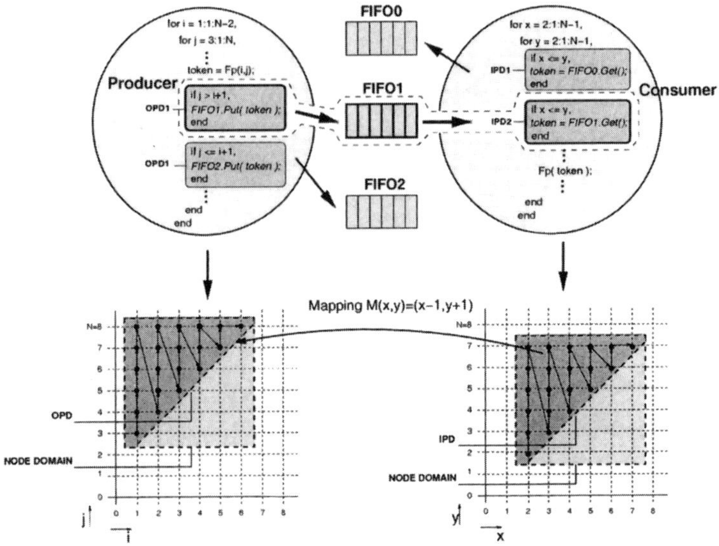

Figure 2 A producer and consumer process. Of the producer we show the output port domains (OPDs) and of the consumer, we show the input port domains (IPDs). Each OPD is uniquely connected to another IPD via a FIFO. Over this FIFO, tokens are communicated that adhere to the mapping given by the mapping matrix M. In this example, OPD1 is connected to IPD2 via FIFO1. The producer/consumer with the FIFO form an instance of the classical consumer/producer pair.

ports in(x,y) defined on the domain $L = \{(x,y) | \, 2 \leq x \leq N \wedge x \leq y \wedge 2 \leq y \leq N\}$. There is a mapping M ($i = x - 1; j = y + 1;$) relating these two port domains. Thus, these two ports form a producer/consumer pair. A token produced by the atomic node p(i,j) is put on the FIFO channel reserved for the edge domain (O, I) through the atomic output port out(i,j), and will be consumed by the atomic node c(x, y) through the atomic input port in(i + 1, j-1) that gets the token from this channel.

Since the KPN processes are sequential processes, no two atomic ports in a port domain are active at the same time. That is, there is an order among the atomic output ports in an output port domain, and there is an order among the atomic input ports in the corresponding input port domain. In [9], we defined the rank function that expresses in a pseudo-polynomial form this order of execution in a particular domain. The *rank* function is derived using the Ehrhart theory that expresses the number of integral points inside of a polytope as a pseudo-polynomial expression [10]. A pseudo-polynomial is a polynomial with periodic coefficients. This theory has been extended recently for parameterized polytopes [11]. As a consequence, the expression of the rank function is, in general, a set of pseudo-polynomial expressions depending on the parameters. Examples of the *rank* functions will be shown later when various realizations of the ELM are discussed.

In Compaan, the sequential ordering of atomic nodes firing in a node domain is in *lexicographical order*, which means these nodes are scheduled according to a loop nest. The ordering in which tokens are put on a channel is the same as the order in which atomic nodes are fired. Because the channel is a FIFO channel, a consumer can only get the tokens from the channel in the same order. This represents the *in-order case*. However, depending on the lexicographical schedule of the consumer's atomic nodes, the consumption of the channel tokens may follow a different order than the order in which these tokens were put on the channel. This represents the *out-of-order* case. To work correctly in the out-of-order case, a consumer needs a mechanism to restore the consumption order. This mechanism relays the use of private *reordering memory* for temporary storage of tokens. Once stored, the tokens can be consumed in the correct order. This reorder mechanism is modeled as the ELM.

III. THE EXTENDED LINEARIZATION MODEL

The main elements in the extended linearization model are the local reordering memory and the controller. Because tokens can no longer be read directly from the FIFO, as they may arrive in the wrong order, they are delivered by the controller to the function unit. In this way, the controller takes care of supplying tokens to the consumer function in the right order. In Fig. 3, a

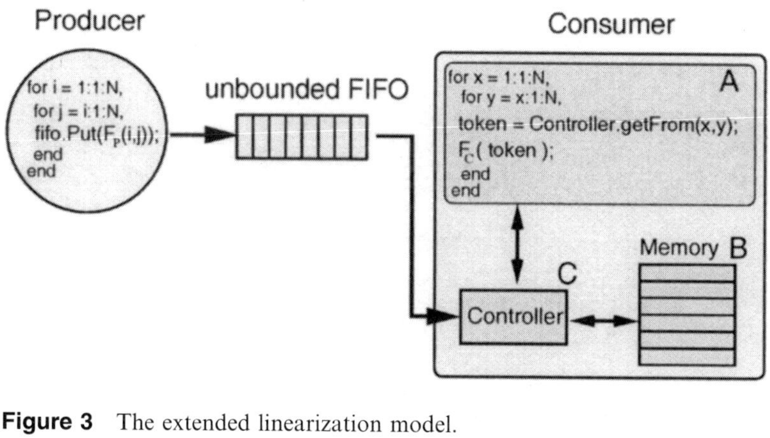

Figure 3 The extended linearization model.

schematic representation is given of the ELM. It shows the consumer process (A), the reorder memory (B), and the controller (C).

A. The Process Description (A)

The process description in the ELM is different from the process description when using the LM. Instead of getting tokens directly from a FIFO, the function gets tokens from the controller. Thus function call *fifo.Get* (see the lower part of Fig. 1) is replaced with the call to the controller function *getFrom*.

B. The Memory (B)

The memory stores tokens allowing the controller to reorder tokens into the order required by the consumer process. Two kinds of memory are possible: random access memory (RAM) and content addressable memory (CAM). The two kinds of memory differ in the way they are addressed. The implementation of the controller depends on the type of the memory.

C. The Controller (C)

The controller converts the sequence tokens are produced into the sequence they have to be consumed. The controller performs this reordering by addressing the reordering memory (B). This functionality is exposed externally to the consumer process by the function *getFrom*(x,y) that returns the token to function F_C for an arbitrary iteration point (x,y).

The behavior of the controller is shown in pseudo code in Fig. 4. The `getReadAddress (x, y)` determines the memory address of the token needed at the iteration (x,y). Next, the controller checks whether the token is already available at that address by calling the function `emptyMem`. If the token is present, it is read from that address by calling the function `read-`

```
Token t getFrom(x,y) {
   double address = getReadAddress(x,y);
   if( ! emptyMem(address) ) {
     return readFromMem(address);
   } else {
     return readFromFifo(address);
   }
}
```

Figure 4 The components in the controller.

`FromMem`. Otherwise, the controller starts to read tokens from the FIFO and stores them in the memory until the desired token arrives at the address of interest. The procedure of reading from FIFO is initiated using the function call `readFromFifo`. Storing tokens into the memory implies that for each token read from the FIFO, a certain address is generated. Depending on the type of memory used, different procedures are available to generate this address. These procedures are realized as the function *getWriteAddress* inside the function `readFromFifo`.

IV. REALIZATIONS OF THE EXTENDED LINEARIZATION MODEL

The ELM can be realized in four different ways as shown in Fig. 5. The realizations differ by the way the function *getReadAddress* and function *getwriteAddress* are implemented and by the type of memory used as re-ordering memory. To compare the four different realizations, the following three characteristics are relevant.

The complexity of the addressing mechanism. The computational complexity of the controller functions *getReadAddress* and *getWrite Address*.

The dimension of the reordering memory. The number of the storage locations needed to perform the reordering.

The generality of the realization. The class of algorithms for which Compaan can derive KPNs.

To introduce the four realizations, we use as an example the producer/consumer pair given in Fig. 6. The graphical representation of the domain descriptions of the producer/consumer pair is shown in the top part of Fig. 7. Because the order in which the producer provides data is different from the order the consumer consumes, an ELM realization is needed in the linearization of the producer/consumer pair.

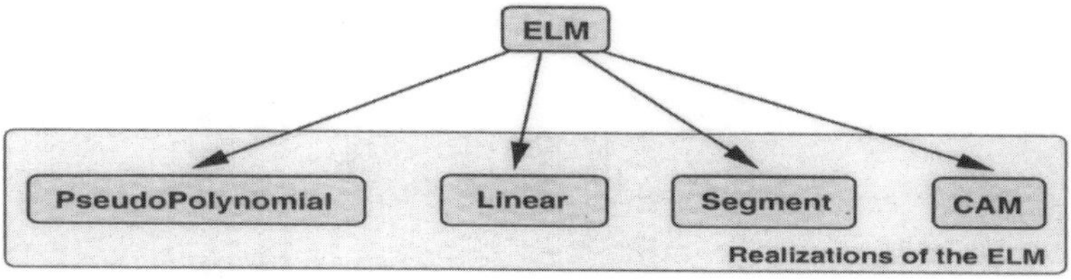

Figure 5 Four model instances.

```
for (int i=1;i<=N+2;i++){
   for (int j=1;j<= N;i++){
      if (2*j >= i+6){
          a[i,j] = Fp();
      }
   }
}
```

Producer

```
for (int y=4;y<= N;y++){
   for (int x=1;x<= N+2;x++){
      if (x <= 2*y-6){
          Fc(a[x,y]);
      }
   }
}
```

Consumer

Figure 6 Running example.

V. PSEUDO-POLYNOMIAL REALIZATION

The pseudo-polynomial realization is based on the fact that the order of the iterations inside an OPD can be expressed as a pseudo-polynomial, which is the *rank* function discussed earlier in this chapter. In general, the *getReadAddress* function of the controller is a pseudo-polynomial function. In Fig. 7, the iteration points of the OPD are perfectly enclosed by a shape that we call the

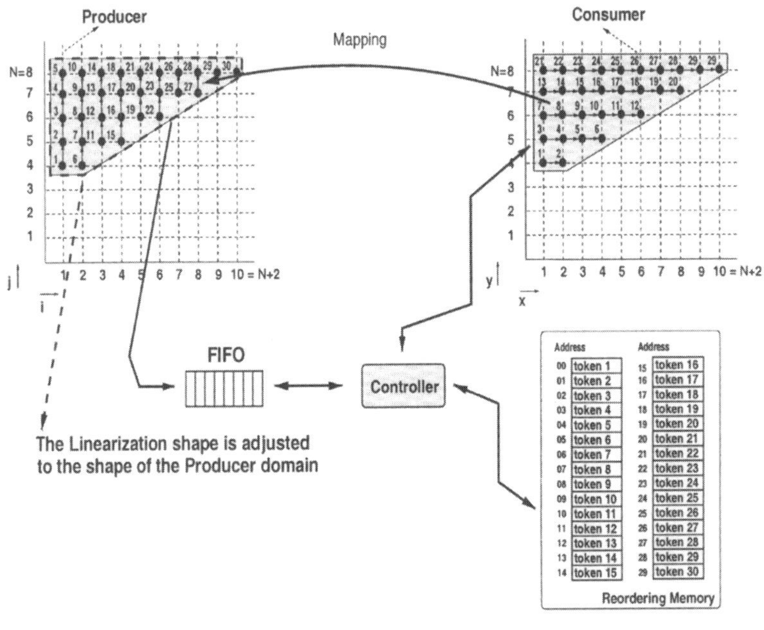

Figure 7 The pseudo-polynomial realization.

linearization shape. The pseudo-polynomial expression is computed by calculating the *rank* function inside the linearization shape. This consists of adding several pseudo-polynomials P_1, $P_2, \ldots$ P_n, where n is equal to the dimension of the linearization shape. For our running example, the rank is the sum of the two pseudo-polynomials P_1 and P_2:

$$\begin{aligned} rank(i,j) &= P_1(i,j) + P_2(i,j) \\ P_1(i,j) &= (i-1) * N + (-1/4) * i^2 - 2 * i + [2, 9/4]_i \\ P_2(i,j) &= j + (-1/2) * i + [-3, -7/2]_i \\ rank(i,j) &= -1/4 * i^2 + (N - 5/2) * i + j + [-1, -5/4]_i - N \end{aligned} \tag{1}$$

To obtain the *getReadAddress* function, the rank needs to be composed with the mapping $M(x,y)$ and the result is equal to:

$$\begin{aligned} getReadAddress(x,y) = rank(i,j) \circ M(x,y) = -1/4 * x^2 \\ + (N - 5/2) * x + y - [1, 5/4]_x - N \end{aligned} \tag{2}$$

The *rank* polynomial contains all the information needed by the Consumer process to reorder the token correctly. For this realization, tokens are written into the reorder memory following the sequence into which they arrive from the FIFO. Therefore, the function *getWriteAddress* is a simple increment. The dimension of the reordering memory is equal to the number of iteration points in the OPD. For the producer/consumer of Fig. 6, the dimension is equal to $(N - 3) * (N + 4)/2$. The computational complexity of addressing the reordering memory can be quite large. It requires the evaluation of a pseudo-polynomial expression like, for example, the one given in equation 2. In general, the pseudo-polynomial realization is valid only for the cases when an OPD is a polytope. Under certain conditions, the realization can be extended for cases an OPD is not a polytope [12].

VI. LINEAR REALIZATION

The linear realization is based on the classical linearization of an n-dimensional array into a one-dimensional array [13,14]. The classical linearization shows that a rectangular shape can be addressed using a simple polynomial. Inspired by this concept, we relax the linearization shape to the smallest rectangular that includes the producer domain (OPD). Consequently, the *getReadAddress* that results is always a simple linear function. The rectangular linearization shape is shown in Fig. 8, and the *rank* function is as follows:

$$rank(i,j) = (N - 3) * (i - 1) + j - 4 \tag{3}$$

The *getReadAddress* function is obtained by composing the *rank* function with the mapping function $M(x,y)$, and the final polynomial expression is

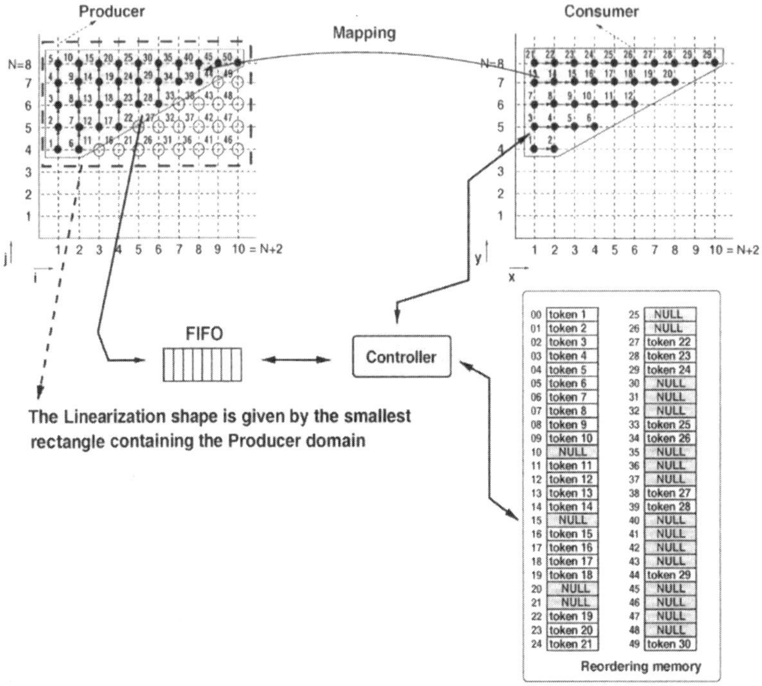

Figure 8 The linear realization.

$$getReadAddress(x, y) = rankp(i,j) \circ M(x, y) \\ = (N - 3) * (x - 1) + y - 4 \tag{4}$$

The consecutive order inside the linearization shape, however, can get disturbed. This happens when an OPD doesn't have a rectangular shape and therefore, more iteration points are enclosed by the linearization shape than necessary. As a consequence, these additional iteration points are also ranked by the *rank* function, disturbing the consecutive order. Looking at Fig. 8, we see that after iteration 10 follows iteration 11. However, iteration 11 does not belong to the OPD. The next iteration belonging to the OPD has rank 12 and thus the order becomes 10, 11, 12. Consequently, the controller cannot rely any longer exclusively on the order tokens are read from the FIFO; the eleventh token read from the FIFO should be written at the address 12. Therefore, the controller cannot use a simple increment for the *getWriteAddress* function.

To recreate the correct sequence of addresses, the controller relays on a function that assigns to incoming tokens the correct order number inside the OPD. This function is called the *recover* function. This function reimplements at the consumer side the logic used to schedule the iteration points inside the OPD.

The advantage of this realization, is that the function used to address the reordering memory is always a linear expression depending on the coordinates of the consumer iteration point. A disadvantage is that need for the *recover* function. Moreover, the extra iteration points enclosed by the linearization shape result in empty memory slots (represented by the "Nulls" in the memory in Fig. 8). In the example, the memory requirement is equal to the dimension of the linearization shape, i.e., to $(N - 3) * (N + 2)$, but only half of this space is actually used.

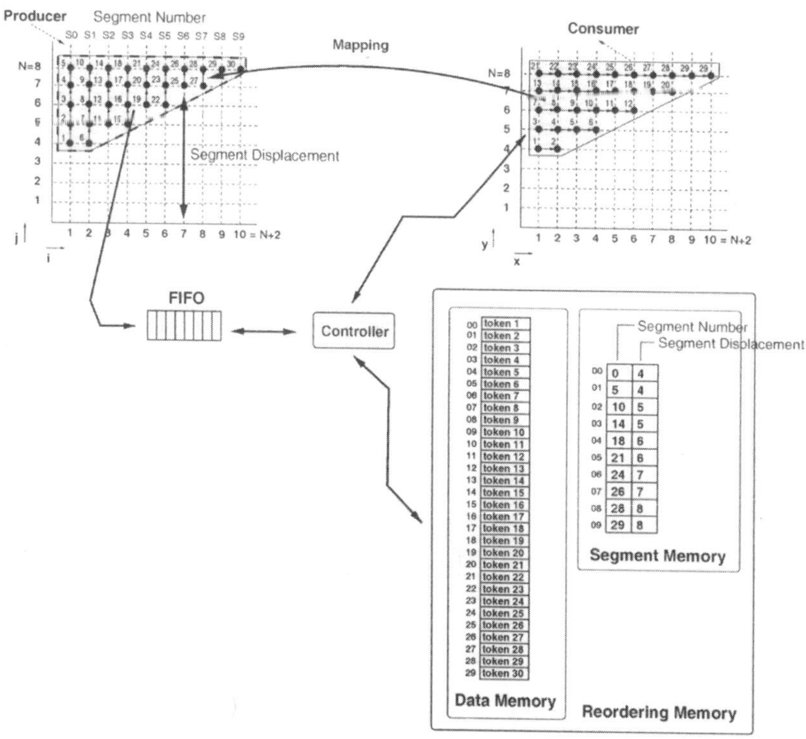

Figure 9 The segment realization.

VII. SEGMENT REALIZATION

The pseudo-polynomial realization results in good memory usage, but the addressing formula can be very complex because of the irregularities it contains as expressed by the periodic coefficients. On the other hand, linear realization results in simple addressing but potentially wastes a lot of memory. We now present the segment realization that combines the best features of the two approaches discussed so far: simple addressing mechanism and efficient memory usage.

The segment realization is based on the fact that pseudo-polynomials can be decomposed into a *linear part* and a *nonlinear part* as shown in Fig. 10. The linear part describes the consecutive order, the nonlinear part described the nonconsecutive order. At the producer side, the order changes at iteration points, at which the innermost nested loops start to iterate again from their lower bound value. We say that a *nonlinearity* occurred at iteration point (IP) and using the notion of these IPs, the segment realization computes the value of the pseudo-polynomial using a *segment number* and a *segment displacement* as is shown in Fig. 10. How the segment number and displacement are computed is explained later in the chapter. Because the segment number and displacement are pre-computed, a pseudo-polynomial can not be evaluated in

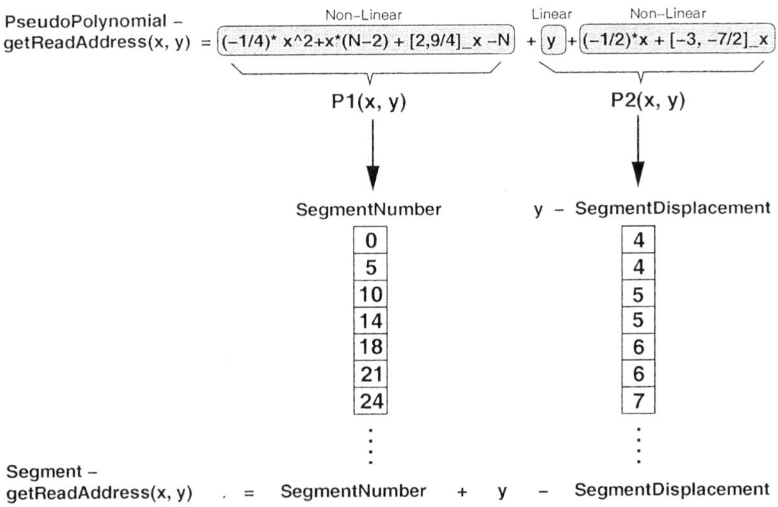

Figure 10 Computation of a pseudo-polynomial using a segment number and a segment displacement stored in the segment memory.

a parameterized way, as is possible in the pseudo-polynomial realization. Thus, parameter values in a NLP must be fixed in order to use the segment realization.

Writing data into the reordering memory occurs in the same way as in the pseudo-polynomial realization. The controller writes tokens in the *reordering memory* as they arrive from the FIFO. The detection of the IPs at which the consecutive order get disturbed, is done by the *recover* function that duplicates the producer for-loops at the consumer side. With each occurrence of an IP, a *segment number* and *segment displacement* is associated. Each such number pair is used by the controller to determine the value of the pseudo-polynomial. Let's see how writing and reading takes place in the segment realization. The writing is implemented in the *getWriteAddress* and the reading is implemented in the *getReadAddress*.

Writing a token happens in the following way. Initially, the controller contains an internal counter that is set to zero. The *recover* function keeps track of whether the order is linear or nonlinear. If the order is linear, a token is read from the FIFO and the internal counter is incremented by one. If the order is non-linear, the controller allocates a new entry in the *segment memory*. It writes in the entry the current value of the iterator in the *segment displacement* field and the currently value of the counter in the segment number field.

In Fig. 9, the producer starts at iteration (1,4), which immediately results in an IP for iterator i. Consequently, an entry is allocated in the segment memory at address 0. The counter has a value of 0 and the iterator is equal to 4, leading to entry (0,4) at address 0. Next, iterator i moves consecutive to iteration (1,8). At the next iteration of i, an IP occurs again. A new entry is generated at address 1. The counter value is equal to 5 and the value of i is again 4, leading to the (5,4) entry at address 1 and so one.

For a particular iteration point of the consumer, the controller determines the address from where data has to be consumed using a three-step procedure. The three steps are

$$\begin{aligned} &\text{Step1.} \quad (i,j) = Map \circ (x,y) \\ &\text{Step2.} \quad Segment = SegmentMemory(i - i_{start}) \\ &\text{Step3.} \quad address = Segment_{Number} + j - Segment_{Displacement} \end{aligned} \tag{5}$$

In Fig. 9, the producer starts from the iteration (i_{start}, j_{start}), which is equal to (1,4). Suppose the consumer wants to obtain the token for iteration (4,6). In Step 1, the iteration is mapped in an iteration at the producer and is equal to (4,6) (i.e., the mapping is identity). In Step 2, the segment is found associated with this iteration. In Step 2, i_{start} is equal to 1 and i is equal to 4. Thus, the segment number is 3, which is address 3 in the segment memory were entry (14,5) is stored. In Step 3, the address in the reordering memory is calculated as $14 + 6 - 5 = 15$. At address 15, the token is stored that was generated at the 16th iteration by the producer. This is the token needed by iteration (4,6) of the consumer as can be verified by inspection in Fig. 9. The memory size of this realization is equal with the size of the data memory plus the size of the segment memory. In our example the size of the datamemory is $N^2/2 + 1/2N - 6$ (the same as the memory size from the pseudo-polynomial realization) and the size of the segmentmemory is $N + 2$. Thus, the total memory size is equal to $N^2/2 + 3/2N - 4$.

VIII. CAM REALIZATION

The content addressable memory (CAM) realization uses CAM as reordering memory. In CAM, a *key* is used instead of an address to access the content of the memory. The entry used in the CAM realization is given in Fig. 11. It shows that each entry in the CAM consists of a key, the token associated with the key, and a field called *multiplicity*. We explain later what the term multiplicity means. The CAM approach works, because to each token produced at the producer OPD an *unique* key can be associated. The controller can reproduce this key to obtain the token the consumer requires at a particular iteration.

For the CAM realization, the function *getReadAddress* generates a key instead of an address. The generation of the unique key can be done in different ways. We compute the *rank* function inside the producer based on a node domain instead of an OPD. Another possibility would have been to use a classical linearization polynomial. In general, the shape of a node domain results in a simple polynomial instead of a pseudo-polynomial and is therefore easily calculated. Using the *rank*, a unique number is associated that is equal to the order of the iteration insidethe node domain of the producer. To

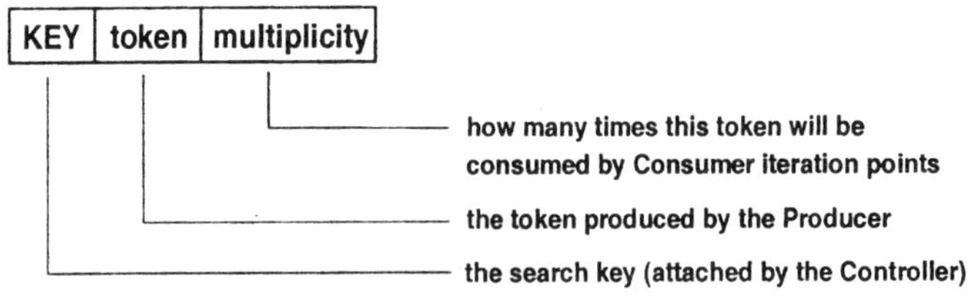

Figure 11 The CAM entry.

illustrate this, consider again the producer/consumer pair from Fig. 6. Suppose that the node domain is defined as

$$C = \{(i,j) | \ 1 \leq i \leq N \wedge 1 \leq j \leq N + 2\}$$

Then the *rank* is given by

$$rank(i,j) = i * N + j$$

By composing the *rank* with the mapping $M(x,y)$ we obtain the *getReadAddress* function:

$$getReadAddress(x, y) = x * N + y \tag{6}$$

For a given iteration point (x,y) of the Consumer process, the *getReadAddress* function determines the unique key for that iteration using eq. 6. For this key, the controller checks (using function *EmptyMem*) if a token already exists in the CAM by searching all keys for a match. If no match can be found, the token is not stored yet.

If the key exists, the token associated with the key is retrieved from the CAM by function *readFromMem*. If the token doesn't exist, the controller keeps loading data from the FIFO into the CAM. This happens in function *ReadFromFifo*. To each token the controller loads, it attaches an unique key given by the function *getWriteAddress* and multiplicity number. Loading data from FIFO stops upon arrival of the token for which the key (as given by *getWriteAddress*) is the same as the key the controller is searching for (as given by *getReadAddress*). The function *getWriteAddress* is based on a *recover* function similarly to the recover function from the linear realization.

In general, a token is read only once by the consumer process. There are cases in which the same token is read more than once by the consumer process. This called a *broadcast*. A read from a FIFO is destructive and in case of a broadcast, this would mean that a token needs to be sent over the FIFO as many times as needed, or that a token needs to be stored in memory and read from memory as many times as needed. In the CAM realization, we implemented the latter option as it is more efficient.

To keep track of how many times a token is to be read, we have introduce the notion of *multiplicity* [12], which indicates how many times a particular token needs to be read by the consumer process. Each time a token is consumed, its multiplicity is decremented. When the multiplicity reaches zero, no other iteration will need that token and it can be erased. That location can be reused by other tokens. Thus, using multiplicity, the controller is able to free memory locations and consequently, this realization uses the smallest possible amount of memory. The memory size (MS) of the CAM is given by the next formula

$$\mathrm{MS} = \max_{(x,y) \in C} (read(x, y) - rank_{Consumer}(x, y)) + 1 \tag{7}$$

where C represents a sub-domain of the consumer domain where no two points read the same token. In the case from Fig. 6, C is the whole consumer domain. The read function is the same as the *getReadAddress* function from the pseudo-polynomial realization:

$$read(x, y) = -1/4 * x^2 + (N - 5/2) * x + y - [1, 5/4]_x - N \tag{8}$$

and the $rank_{Consumer}$ is the function that gives the order of the consumer iteration points:

$$rank_{Consumer}(x, y) = y^2 - 7y + x + 11 \tag{9}$$

According to Eqs. 8 and 9 it results with

$$\text{MS} = \max_{(x,y)\in C}(-1/4 * x^2 + (N - 7/2)x - y^2 + 8y - N - [11, 45/4]_x)$$

The maximum of this formula inside the C domain can be derived using analitycal methods. In our case for $N = 8$, we have $MS = 10$. For more informations about the read and the *rank* function, we refer to [9]. The key to efficient memory usage is the ability to compute the multiplicity for a token. However, this multiplicity is again computed using the Ehrhart theory and may again be a pseudo-polynomial.

IX. COMPARING THE DIFFERENT REALIZATIONS

In the previous sections, we have shown four different realizations for ELM. Each realization has its strength in terms of efficiency of memory usage and computational complexity of address the memory. In this section we offer general remarks about the different realizations and summarize the strengths and weaknesses.

A. General Remarks

1. Linearization Shapes

In the realizations, the linearization shape of the OPD determines the complexity of the *getReadAddress* implementation. We indicated that when rectangles are enforced, simple polynomials result. There are application domains where the rectangular shape is the natural linearization shape, for example, in imaging. In those cases, the linear realization does not have the disadvantage of memory wastes and the need for a *recover* function. On the other hand, we found more complex linearization shapes in advanced signal processing algorithms. In algorithms like QR or SVD, triangular shapes typically lead to complex pseudo-polynomials.

2. Parameterized Versus Static Realizations

We solved the linearization under the assumption that we want to keep the problem parameterized in the original parameters of the loo-bounds of the parameterized NLPs. If, however, we need to provide a realization for specific values of these parameters, we can use a much simpler realization of the controller. The segment realization already shows that. In general, if we can evaluate the *getReadAddress* a priori, the controller becomes a simple look-up table.

3. RAM versus CAM

RAM is the most commonly used form of memory. It is simple, cheap, and widely available on FPGAs. But more and more, CAMs are also becoming available. Today, there are FPGA platforms available on the market that supports CAM blocks with high speed search time [15].

4. Dense Polytopes

In the examples shown so far, we assumed that all nodes in the OPD and IPD can be enclosed by a linearization shape. There are cases, however, in which we can find the exact shape, but still not all points are part of the enclosure. We refer to these points as *holes* and they are introduced when a four-loop is used with a stride other than one, or when linear expresions are used that contain operators like mod, div, floor, ceil, max, or min. The holes affet the generality of the realizations presented in this chapter. Not all of the discussed realizations can handle holes. For example, if a linearization shape encloses holes, these holes also get ranked, thereby disturbing the consecutive order.

5. Recover Function

In three of the presented realizations, the function *getWriteAddress* is based on the *recover* function. For each token read from the FIFO, this function recovers the iteration at which this token was produced inside the IPD. Basically such a function duplicates the control from the IPD as a finite state machine, which can be computationally expensive.

Instead of using the *recover* function at the consumer, another approach would be to tag the tokens produced at the OPD with additional information. In this way, the controller and memory have the same function as the *matching unit* found in classical Dataflow architectures [16]. The problem in these matching units was to find a lower bound on its memory, such that a program would not deadlock. We have shown in Eq. 7 that we can determine a lower bound on the memory such that no deadlock occurs given the class of parameterized NLPs.

6. Practical Limitations

The pseudo-polynomial realization depends very much on the ability to calculate the *rank* functions. We rely on the polylib library [17], to compute the *rank* function. Although this library has proven to be quite stable and useful, this implementation of the Ehrhart theory is not always able to compute the *rank* function. By selecting the linear, segment, or CAM realization, we are always able to come up with a representation of a KPN.

B. Summary

We have presented four different realizations for ELM. In Fig. 12, we compare these realizations for the producer/consumer given in Table 1. The table shows the memory requirements in a symbolic way for parameter N and when $N = 8$, the computational complexity of addressing the memory (as done by function *getReadAddress*), whether a *recover* function is needed, and finally the generality of the approach. We can see that the segment realization uses more memory than the pseudo-polynomial realization because the segment part consumes some memory. The advantage of the segment realization is that the controller can fill the memory in the same order tokens arrive. The segment realization uses less memory than the linear realization. The computational complexity of addressings, in the segment or pseudo-polynomial cases, is less for the segment realization although in both cases a pseudo-polynomial is evaluated. Finally, we observe that the CAM realiza-

Table 1 Comparison of ELM realizations

Linearization model	Memory size	$N=8$	Computational complexity	Recover	Generality
Pseudo	$N^2/2 + 1/2 \cdot N - 6$	30	$C1$	No	No
Linear	$N^2 - N - 6$	50	$C2 \ll C1$	Yes	Yes
Segment	$N^2/2 + 3/2 \cdot N - 4$	40	$C3 \sim C2$	Yes	Yes
CAM	$max(read - rank)$	10	$C4 \sim C2$	Yes	Yes

tion uses the least amount of memory but may require a relatively complex addressing mechanism since the CAM realization requires the computation of unique keys.

X. CONCLUSIONS

In this chapter, we presented the extended linearization model (ELM). This model was introduced to solve out-of-order communications of tokens between a producer and a consumer process. The ELM adds to a process some additional memory and a controller without violating the Kahn Process Network semantics; we still use a FIFO between a producer and consumer. The controller uses the local memory to reorder the tokens in the order the consumer expects the tokens. In the realization of the ELM, the implementation of the controller is the difficult part.

To implement the controller, we make a lot of use of the *rank* function. This function assigns to an arbitrary iteration a unique rank number that indicates when it is produced. The *rank* function is in general a pseudo-polynomial. We exploit the ability to derive such polynomial at compile time to find realizations for the ELM at compile time. In the realizations, we assumed that we only exchange tokens over a FIFO without any additional information. We did not assume tagging of tokens.

When realizing the ELM, we have seen that four different realizations exist. The first realization is the pseudo-polynomial realization. It uses exclusively pseudo-polynomials to solve the reordering case. The advantage is that we can sove the reordering in a parameterized way. Because in the most general case pseudo-polynomials are involved, the implementation can be computationally complex. Also, the pseudo-polynomial can, in practice, not always be calculated. If we relax the linearization shape to the smallest rectangular that encloses all iterations, we obtain a more simple implementation. However, this might be at the expense of inefficient memory usage. If we want to avoid complex addressing and efficient memory usage, the segment realization is a good choice, although the solution is no longer parameterized. Finally, we showed that we can use a key instead of an address to retrieve the proper tokens. The calculation of this key is, in general, a simple polynomial that is easy to realize. Also, the CAM realization requires the least amount of memory to solve the reordering of tokens.

The KNPs derived by Compaan can be simulated using the YAPI framework [18] or by using the PN-domain in Ptolemy II [19]. In both cases, we must implement the presented realizations in software. We are currently able to implement in software, at compile time, the pseudo-polynomial, and

CAM realization. For these realizations, we have shown that we can derive correct implementations. We verified this by running Compaan on a set of applications written as parameterized NLPs.

REFERENCES

1. Kienhuis, B., Deprettere, E., Vissers, K., van der Wolf, P. (1997). An approach for quantitative analysis of application-specific dataflow architectures. In: Proceedings of 11th Int. Conference of Applications-specific Systems, Architectures and Processors (ASAP'97), Zurich, Switzerland, pp. 338–349.
2. Kienhuis, B., Deprettere, E., van der Wolf, P., Vissers, K. (2002). A Methodology to Design Programmable Embedded Systems, Volume 2268 of LNCS. Springer Verlag, pp. 18–37.
3. Kienhuis, B., Rijpkema, E., Deprettere, E.F. (2000). Compaan: Deriving process networks from matlab for embedded signal processing architectures. In: 8th International Workshop on Hardware/Software Codesign (CODES'2000), San Diego, USA.
4. Wolf, W. (2001). Computers as Components—Principles of Embedded Computing System Design. Basel: Morgan Kaufmann Publishers, Inc.
5. Lee, E.A., Parks, T.M. (1995). Dataflow process networks. Proceedings of the IEEE 83:773–799.
6. Kahn, G. (1974). The semantics of a simple language for parallel programming. Proc. of the IFIP Congress 74, North-Holland Publishing Co.
7. Harriss, T., Walke, R., Kienhuis, B., Depettere, E. (2002). Compilation from matlab to process networks realized in fpga. Design Automation of Embedded Systems 7.
8. Ben-Ari, M. (1982). Principles of Concurrent Programming. Prentice Hall.
9. Turjan, A., Kienhuis, B., Deprettere, E. (2002). A compile time based approach for solving out-of-order communication in kahn process networks. In: Proceedings of the IEEE 13th International Conference on Application-specific Systems, Architectures and Processors (ASAP'02), San Jose, California.
10. Ehrhart, E. (1977). Polynômes arithmétiques et Méthode des Polyédres en Combinatoire. International series of numerical mathematics. Vol. 35 edn. Basel: Birkhäuser Verlag.
11. Clauss P., Loechner V. (1998). Parametric analysis of polyhedral iteration spaces. Journal of VLSI Signal Processing 19:179–194.
12. Rijpkema, E. (2002). Modeling Task Level Parallelism in Piece-wise Regular Programs. PhD thesis, University Leiden, LIACS, Leiden, The Netherlands.
13. Aho, A.V., Sethi, R., Ullman, J.D. (1986). Compilers: Principles, Techniques and Tools. Addison-Wesley ISBN 0-201-10088-6.
14. Muchnick S. (1997). Advanced Compiler Design and Implementation. Morgan Kaufmann Publishers, Inc.
15. Xilinx (2001). Memory Application Notes for Virtex-II Devices. http://www.xilinx.com, pp. 505–506.

16. Veen, A.H. (1986). Dataflow machine architecture. ACM Computing Surveys 18:366–396.
17. Clauss, P., Loechner, V. (2002). *Polylib* (http://icps.u-strabg.fr/Polylib).
18. de Kock, E., Essink, G., Smits, W., van der Wolf, P., Brunel J., Kruijtzer, W., Lieverse, P., Vissers, K. (2000). Yapi: Application modeling for signal processing systems. In: 37th Design Automation Conference, Los Angeles, CA.
19. Davis, J., II, Hylands, C., Kienhuis, B., Lee, E.A., Liu, J., Liu, X., Muliadi, L., Neuendorffer, S., Tsay, J., Vogel, B., Xiong, Y. (2000). Ptolemy ii - heterogeneous concurrent modeling and design in java.

10

Communication Services for Networks on Chip

Andrei Rădulescu and Kees Goossens
Philips Research Laboratories, Eindhoven, The Netherlands

I. INTRODUCTION

Networks on chip (NoC) have received considerable attention recently as a solution to the interconnect problem in highly-complex chips [3–5,7–9,15, 20,23]. The reason is two-fold. First, NoCs help resolve the electrical problems in new deep-submicron technologies, as they structure and manage global wires [3–5,7,8]. At the same time they share wires, lowering their number and increasing their utilization [7,8]. NoCs can also be energy efficient and reliable [4], and are scalable compared to buses [9]. Second, NoCs also decouple computation from communication, which is essential in managing the design of billion-transistor chips [14,23] NoCs achieve this decoupling because they are traditionally designed using protocol stacks [22], which provide well-defined interfaces separating communication service usage from service implementation [5,23].

Using networks for on-chip communication when designing systems on chip (SoC), however, raises a number of new issues that must be taken into account. This is because, in contrast to existing on-chip interconnects (e.g., buses, switches, or point-to-point wires), where the communicating modules are directly connected, in a NoC the modules communicate remotely via network nodes. As a result, interconnect arbitration changes from centralized to distributed, and issues like out-of order transactions, higher latencies, and end-to-end flow control must be handled either by the intellectual property (IP) block or by the network.

Most of these topics have already been the subject of research in the field of computer networks [25] and parallel machine interconnect networks [6]. However, on-chip networks have different properties (e.g., tighter link synchronization) and constraints (e.g., higher memory cost) leading to different design choices, which ultimately affect the network services.

In this chapter, we compare NoCs and off-chip networks showing both their similarities and differences. We also explore the differences between NoCs and existing on-chip interconnects. We present an interface that takes these issues into consideration. Our interface is aimed at being similar to a split-transaction bus interface, such as VCI [26], OCP [17] or DTL [18], to allow simple, low-cost wrappers to bus interfaces, and to allow backward compatibility with existing IPs. Our interface uses a request-response protocol that provides basic read and write operations. But it extends bus interfaces to fully exploit the power of our NoC [8,20,21]. For example, it offers connection-based communication where end-to-end flow control and time-related guarantees (e.g., bounded latency) can be requested.

The chapter is organized as follows. In the next two sections we compare NoC properties with those of off-chip networks and buses, respectively. In Section IV, we define the services that we offer in our network. Finally, we present our conclusions.

II. NETWORKS BROUGHT ON CHIP

Networks have been the subject of research for decades, both in the context of local and wide-area networks (computer networks) [25], and as an interconnect for parallel machines [6]. Both are very much related to on-chip networks, and many of the results in those fields are also applicable on chip. However, NoC premises are different from off-chip networks and, therefore, most of the network design choices must be re-evaluated.

NoCs differ from off-chip networks mainly in their constraints and synchronization. Typically, resource constraints are tighter on chip than off chip. Storage (i.e., memory) and computation resources are relatively more expensive, whereas the number of point-to-point links is larger on chip than off chip [7].

Storage is expensive, because general-purpose on-chip memory, such as RAMs, occupy a large area. Having the memory distributed in the network components in relatively small sizes is even worse, as the overhead area in the memory then becomes dominant.

For on-chip networks, computation also comes at a relatively high cost compared to off-chip networks. An off-chip network interface usually contains a dedicated processor to implement the part of the protocol stack to relieve the host processor from the communication processing. Including a dedicated processor in a network interface may be not feasible on chip, as the size of the network interface will become comparable to, or larger than, the IP to be connected to the network. Moreover, running the protocol stack on the IP itself may also not be feasible, because often these IPs have one dedicated function only, and do not have the capabilities to run a network protocol stack. A cost-effective solution would be to have a dedicated-hardware implementation of the protocol stack.

The number of wires and pins to connect network components is an order of magnitude larger on chip than off chip [7]. If they are not used massively for other purposes than NoC communication, they allow wide point-to-point interconnects (e.g., 300-bit links) [7,15]. This is not possible off-chip, where links are relatively narrower: 8–16 bits.

On-chip wires are also relatively shorter than off chip [7], allowing a much tighter synchronization than off chip. This allows a reduction in the buffer space in the routers because the communication can be done at a smaller granularity. In the current semiconductor technologies, wires are also fast and reliable, which allows simpler link–layer protocols (e.g., no need for error correction, or retransmission). This also compensates for the lack of memory and computational resources.

In the rest of this section, we list five network issues that have a direct impact on the NoC cost: reliable communication, deadlock, data ordering, network flow control and buffering strategy, and time-related guarantees. For each of them, we discuss the differences and similarities for on- and off-chip networks.

1. Reliable Communication

A consequence of the tight on-chip resource constraints is that the network components (i.e., routers and network interfaces) must be fairly simple to minimize computation and memory requirements. Luckily, on-chip wires currently provide a reliable communication medium, which can help to avoid the considerable overhead incurred by off-chip networks for providing reliable communication. Data integrity can be provided at low cost at the data link layer. However, data loss also depends on the network architecture. In most computer networks data is simply dropped if congestion occurs in the network [6,25]. On-chip, dropping data may lead to a too costly implementation of reliable communication. We show below that a network that does not drop data can be a much lower-cost solution, at the peril of introducing the possibility of deadlock.

2. Deadlock

Computer network topologies have generally an irregular (possibly dynamic) structure, which can introduce buffer cycles. In such topologies, packet dropping at the network nodes may be required to avoid deadlocks.

Deadlock can also be avoided without dropping data, for example by introducing constraints either in the topology or routing. Fat-tree topologies have already been considered for NoCs, where deadlock is avoided by bouncing back packets in the network in case of buffer overflow [9]. Tile-based approaches to system design [7,15,24] use mesh or torus network topologies, where deadlock can be avoided using, for example, a turn-model routing algorithm [6].

An alternative solution for deadlock in NoCs, which takes into consideration that modules connecting to the network are either masters (initiating requests and receiving responses), or slaves (receiving requests and sending back responses), is to maintain separate virtual networks (with separate buffers) for requests and responses [6].

3. Data Ordering

In a network, data sent from a source to a destination may arrive out of order due to reordering in network nodes, following different routes, or retransmission after dropping. For off-chip networks out-of-order data delivery is typical. However, for NoCs where no data is dropped, data can be forced to follow the same path between a source and a destination (deterministic routing) with no reordering. This in-order data transportation requires less buffer space, and reordering modules are no longer necessary.

4. Network Flow Control and Buffering Strategy

Network flow control and buffering strategy have a direct impact on the memory utilization in the network. Wormhole routing requires only a flit buffer (per queue) in the router, whereas store-and-forward and virtual-cut-through routing require at least the buffer space to accommodate a packet [6]. Consequently, on-chip, wormhole routing may be preferred over virtual-cut-through or store-and-forward routing. Similarly, input queuing may be a lower memory cost alternative to virtual-output-queuing or output-queuing buffering strategies, because it has fewer queues. Dedicated (lower cost) FIFO memory structures also enable on-chip usage of virtual-cut-through routing or virtual-output-queuing for a better performance [20]. However, using virtual-cut-through routing and virtual-output-queuing at the same time is still too costly [20].

5. Time-Related Guarantees

Off-chip networks typically use packet switching and offer best-effort services. Contention can occur at each network node, making latency guarantees very hard to offer. Throughput guarantees can still be offered using schemes such as rate-based switching [27] or deadline-based packet switching [19], but with high buffering costs.

An alternative to provide such time-related guarantees is to use time-division multiple access (TDMA) circuits, where every circuit is dedicated to a network connection. Circuits provide guarantees at a relatively low memory and computation cost. Network resource utilization is increased when the network architecture allows any leftover guaranteed bandwidth to be used by best-effort communication [10,20,21].

III. FROM BUSES TO NOCs

Introducing networks (Fig. 1) as on-chip interconnects radically changes the communication as compared to direct interconnects such as buses or switches (Fig. 2). This is because of the multi-hop nature of a network, where communication modules are not directly connected, but separated by one or more network nodes. This is in contrast with the prevalent existing interconnects (i.e., buses) where modules are directly connected. The implications of this change reside in the arbitration (which must change from centralized to distributed), and in the communication properties (e.g., ordering, or flow control).

In this section, we list some of these topics, and outline the differences between NoCs and buses. We refer mainly to buses as direct interconnects because currently they are the most used on-chip interconnect. Most of the

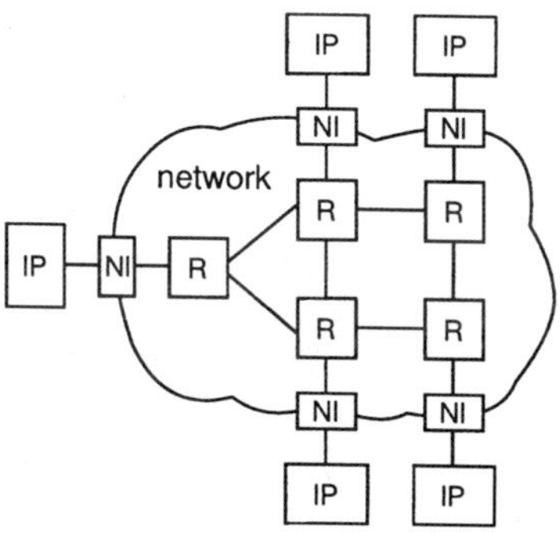

Figure 1 A network interconnect example.

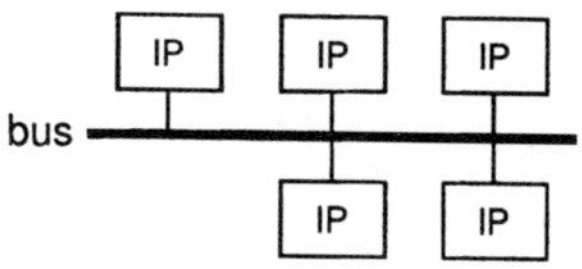

Figure 2 A bus interconnect example.

bus characteristics also hold for other direct interconnects (e.g., switches [16]). Multilevel buses are a hybrid between buses and NoCs. For our purposes, depending on the functionality of the bridges, multilevel buses behave either like simple buses [2] or like NoCs.

1. Programming Model

The programming model of a bus typically consists of load and store operations which are implemented as a sequence of primitive bus transactions. Bus interfaces typically have dedicated groups of wires for command, address, write data, and read data [1,12,13,17,18,26].

A bus is a resource shared by multiple IPs. Therefore, before using it, IPs must go through an arbitration phase, where they request access to the bus, and block until the bus is granted to them.

A bus transaction involves a request and possibly a response. Modules issuing requests are called masters, and those serving requests are called slaves. If there is a single arbitration for a request–response pair, the bus is called nonsplit. In this case, the bus remains allocated to the master of the transaction until the response is delivered, even when this takes a long time. Alternatively, in a split bus, the bus is released after the request to allow transactions from different masters to be initiated. However, a new arbitration must be performed for the response such that the slave can access the bus [11].

For both split and nonsplit buses, both communication parties have direct and immediate access to the status of the transaction. In contrast, network transactions are one-way transfers from an output buffer at the source to an input buffer at the destination that causes some action at the destination, the occurrence of which is not visible at the source [6]. The effects of a network transaction are observable only through additional transactions. A request–response type of operation is still possible, but requires at least two distinct network transactions. Thus, a bus-like transaction in a NoC will essentially be a split transaction.

2. Transaction Ordering

Traditionally, all transactions on a bus are ordered (Peripheral VCI [26], AMBA [1], DTL [18], or CoreConnect PLB and OPB [12, 13]). This is possible at low cost, because the interconnect, being a direct link between the communicating parties, does not reorder data. However, on a split bus, a total ordering of transactions on a single master may still cause performance penalties, when slaves respond at different speeds. To solve this problem, recent extensions to bus protocols allow transactions to be performed on connections. Ordering of transactions within a connection is still preserved, but between connections there are no ordering constraints (e.g., OCP [17], or Basic VCI

[26]). A few of the bus protocols allow out-of-order responses per connection in their advanced modes (e.g., Advanced VCI [26]), but both requests and responses arrive at the destination in the same order as sent.

In a NoC, ordering becomes weaker. Global ordering can only be provided at a very high cost due to the conflict between the distributed nature of the networks, and the requirement of a centralized arbitration necessary for global ordering.

Even local ordering, between a source–destination pair, may be costly. Data may arrive out of order if it is transported over multiple routes. In such cases, to still achieve an in-order delivery, data must be labeled with sequence numbers and reordered at the destination before being delivered.

3. Atomic Chains of Transactions

An atomic chain of transactions is a sequence of transactions initiated by a single master that is executed on a single slave exclusively. That is, other masters are denied access to that slave, once the first transaction in the chain claimed it. This mechanism is widely used to implement synchronization mechanisms between master modules (e.g., semaphores).

On a bus, atomic operations can easily be implemented, as the central arbiter will either lock the bus for exclusive use by the master requesting the atomic chain, or not grant access to a locked slave. In the former case, the duration the resources are locked is shorter because once a master has been granted access to a bus, it can quickly perform all the transactions in the chain (no arbitration delay is required for the subsequent transactions in the chain). Consequently, the locked slave and the bus can be opened again in a short time. This approach is used in AMBA and CoreConnect. In the latter case, the bus is not locked, and can still be used by other modules, however, at the price of a longer locking duration of the slave. This approached is used in VCI and OCP.

In a NoC, where the arbitration is distributed, masters do not know that a slave is locked. Therefore, transactions to a locked slave may still be initiated, even though the locked slave cannot accept them. Consequently, to prevent deadlock, these other transactions must be either dropped, or transactions in the atomic chain must be able to bypass them to be served. Moreover, the duration a module is kept locked is much longer in case of NoCs, because of the higher latency per transaction.

4. Deadlock

In buses, deadlock is generally not an issue. Deadlock can still occur at the application level (e.g., an atomic chain of transactions that locks the bus, but never unlocks it), but this is not caused by the interconnect itself.

In a network, deadlock becomes a more important issue, and special care must be taken in the network design to avoid deadlock. Deadlock is mainly caused by cycles in the buffers. To avoid deadlock, either network nodes must drop packets when their buffers are filled, or routing must be cycle-free. In a NoC, we believe the latter is preferable, because of its lower cost in achieving reliable communication (see Section II).

A second cause of deadlock is atomic chains of transactions. The reason is that while a module is locked, the storing transactions of the queue get filled with transactions outside the atomic transaction chain, blocking access of the transaction in the chain to reach the locked module. If atomic transaction chains must be implemented (to be compatible with processors allowing this, such as MIPS), the network nodes should be able to filter the transactions in the atomic chain, or be allowed to drop those blocking them.

5. Media Arbitration

An important difference between buses and NoCs is the medium arbitration scheme. In a bus, master modules request access to the interconnect and the arbiter grants access for the whole interconnect at once. Arbitration is *centralized* as there is only one arbiter component. It is also *global* as all the requests as well as the state of the interconnect are visible to the arbiter. Moreover, when a grant is given, the complete path from the source to the destination is exclusively reserved.

In a nonsplit bus, arbitration takes place once when a transaction is initiated. As a result, the bus is granted for both request and response. In a split bus, requests and responses are arbitrated separately.

In a NoC arbitration is also necessary, as it is a shared interconnect. However, in contrast to buses, the arbitration is *distributed*, because it is performed in every router, and is based only on local information. Arbitration of the communication resources (links, buffers) is performed incrementally as the request or response advances [20].

6. Destination Name and Routing

For a bus, the command, address, and data are broadcasted on the interconnect. They arrive at every destination, only one of which activates, based on the broadcasted address, and executes the requested command. This is possible because all modules are directly connected to the same bus.

In a NoC, it is not feasible to broadcast information to all destinations, because it must be copied to all routers and network interfaces. This floods the network with data. The address is better decoded at the source to find a route to the destination module. A transaction address has, therefore, two parts: (1) a destination identifier, and (2) an internal address at the destination.

7. Latency

Transaction latency is caused by two factors: (1) the access time to the bus, which is the time until the bus is granted, and (2) the latency introduced by the interconnect to transfer the data.

For a bus, where the arbitration is centralized, the access time is proportional to the number of masters connected to the bus. The transfer latency itself typically is constant and relatively low, because the modules are linked directly. However, the speed of transfer is limited by the bus speed, which is relatively low.

In a NoC, arbitration is performed at each router for the following link. The access time per router is short. Both end-to-end access time and transport time increase proportionally to the number of hops between master and slave. However, network links are unidirectional and point to point and, thus can run at higher frequencies than buses, thus lowering the latency.

From a latency prospective, using a bus or a network is a trade-off between the number of modules connected to the interconnect (which affects access time), the speed of the interconnect, and the network topology.

8. Data Format

In most modern bus interfaces the data format is defined by separate wire groups for the transaction type, address, write data, read data, and return acknowledgments/errors (e.g., VCI, OCP, AMBA, DTL, or CoreConnect). This is used to pipeline transactions. For example, concurrently with sending the address of a read transaction, the data of a previous write transaction can be sent, and the data from an even earlier read transaction can be received. Moreover, having dedicated wire groups simplifies the transaction decoding; there is no need for a mechanism to select between different kinds of data sent over a common set of wires.

Inside a network, there is typically no distinction between different kinds of data. Data is treated uniformly and passed from one router to another. This is done to minimize the control overhead and buffering in routers. If separate wires would be used for each of the above-mentioned groups, separate routing, scheduling and queuing would be needed, and the cost of routers would increase proportionally.

In addition, in a network at each layer in the protocol stack, control information must be supplied together with the data (e.g., packet type, network address, or packet size). This control information is organized as an envelope around the data. That is, first a header is sent, followed by the actual data (payload), followed possibly by a trailer. Multiple envelopes may be provided for the same data with each carrying the corresponding control information for each layer in the network protocol stack [6,25].

9. Buffering and Flow Control

Buffering data of a master (output buffering) is used both for buses and NoCs to decouple computation from communication. However, for NoCs output buffering is also needed to marshal data, which consists of (1) (optionally) splitting the outgoing data into smaller packets that are transported by the network, and (2) adding control information for the network around the data (packet header). To avoid output buffer overflow the master must not initiate transactions that generate more data than the currently available space.

Similarly to output buffering, input buffering is also used to decouple computation from communication. In a NoC, input buffering is also required to unmarshal data.

In addition, flow control for input buffers differs for buses and NoCs. For buses, the source and destination are directly linked and destination can, therefore, signal directly to a source that it cannot accept data. This information can even be available to the arbiter such that the bus is not granted to a transaction trying to write to a full buffer.

In a NoC, however, the destination of a transaction cannot signal directly to a source that its input buffer is full. Consequently, transactions to a destination can be started, possibly from multiple sources, after the destination's input buffer has filled up. Several policies can be adopted when an input buffer is full. One policy is not to accept additional incoming transitions and to store them in the network. However, this approach can easily lead to network congestion, as the data could eventually be stored all the way to the sources, blocking the links in between. Another policy is to accept incoming transactions at a full destination and drop some data in the input buffer. Congestion is avoided but data is lost and this is undesirable.

To avoid input buffer overflow, connections can be used together with end-to-end flow control. At connection set up between a master and one or more slaves, buffer space is allocated at the network interfaces of the slaves, and the network interface of the master is assigned credits reflecting the amount of buffer space at the slaves. The master can only send data when it has enough credits for the destination slave(s). The slaves grant credits to the master when they consume data.

IV. THE ÆTHEREAL APPROACH

As described in the previous two sections, NoCs have different properties from both existing off-chip networks and existing on-chip interconnects. As a result, existing protocols and service interfaces cannot be adopted directly to NoCs, but must take the characteristics of NoCs into account. For example, a protocol such as TCP/IP assumes the network is loss and includes significant complexity to provide reliable communication. Therefore, TCP/IP is not suitable in a NoC where we assume data transfer reliability is already solved at a lower level. On the other hand, existing on-chip protocols such as VCI, OCP, AMBA, DTL, or CoreConnect are also not directly applicable. For example, they assume ordered transport of data: if two requests are initiated from the same master, they will arrive in the same order at the destination. This does not hold automatically for NoCs. Atomic chains of transactions and end-to-end flow control also need special attention in a NoC interface.

Our objectives when defining the services of our on-chip network (called Æthereal) are the following. First, the services abstract from the network internals as much as possible. This is a key ingredient in tackling the challenge of decoupling the computation from communication [14,23], which allows IPs (the computation part), and the interconnect (the communication part) to be designed independently from each other. As a consequence, our services are positioned at the transport layer in the ISO-OSI reference model [25], which is the first layer to be independent of the implementation of the network.

Second, we aim at a NoC interface as close as possible to a bus interface. NoCs can then be introduced nondisruptively: existing IPs, methodologies, and tools can continue to be used with minor changes. As a consequence, we use a request–response interface similar to interfaces for split buses [1,12,13, 17,18,26].

Third, our interface extends traditional bus interfaces to fully exploit the power of NoCs. For example, we offer connection-based communication, which does not only relax ordering constraints (as for buses), but also enables new communication properties such as end-to-end flow control based on credits, or guaranteed throughput [8,20,21]. All these properties can be set for each connection individually.

A. The Æthereal Connection and Transaction Model

IPs interact with our network [8,20,21] at so-called network interfaces (NI). NIs provide network interface parts (NIPs) through which the communication services are accessed. As shown in Fig. 3, a NI can have several NIPs, to which one or more IPs (computation elements or memories, but not interconnection elements) can be connected. Similarly, an IP can be connected to more than one NI and NIP.

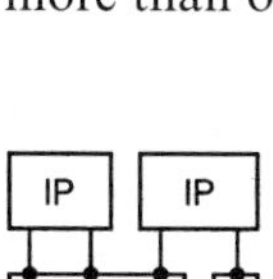

Figure 3 Examples of links between NIs and IPs.

Communication between NIPs is performed on *connections*. Connections are introduced to describe and identify communication with different properties, such as guaranteed throughput, bounded latency and jitter, ordered delivery, or flow control. For example, to distinguish and independently guarantee communication of 1Mbs and 25Mbs, two connections can be used. Two NIPs can be connected by multiple connections, possibly with different properties. Connections (as defined here) are similar to the concept of threads and connections from OCP and VCI. Where OCP and VCI connections are used only to relax transaction ordering, we generalize from only the ordering property to include configuration of buffering and flow control, guaranteed throughput, and bounded latency per connection.

Æthereal connections must be *created* with the desired properties before being used. This may result in resource reservations inside the network (e.g., buffer space, or percentage of the link usage per time unit). If the requested resources are not available, the network will refuse the request. After usage, connections are *closed*, which leads to freeing the resources occupied by that connection.

To allow more flexibility in configuring connections and, thus better resource allocation per connection, the outgoing and return parts of connections are configured independently. For example, a different amount of buffer space can be allocated in the NIPs at master and slaves, or different bandwidths can be reserved for requests and responses.

Depending on the requested services, the time to handle a connection (i.e., creating, closing, modifying services) can be short (e.g., creating/closing an unordered, lossy, best-effort connection) or significant (e.g., creating/closing a multicast guaranteed-throughput connection). Consequently, connections are assumed to be created, closed, or modified infrequently, coinciding (e.g., with reconfiguration points) when the application requirements change.

Communication takes place on connections using *transactions*, consisting of a request and possibly a response. The request encodes an operation (e.g., read, write, flush, test and set, nop) and possibly carries outgoing data (e.g., for write commands). The response returns data as a result of a command (e.g., read) and/or an acknowledgment.

Connections involve at least two NIPs. Transactions on a connection are always started at one and only one of the NIPs, called the connection's *active* NIP (ANIP). All the other NIPs of the connection are called *passive* NIPs (PNIP).

There can be multiple transactions active on a connection at a time, but more generally than for split buses. That is, transactions can be started at the ANIP of a connection while responses for earlier transactions are pending. If a connection has multiple slaves, multiple transactions can be initiated towards different slaves. Transactions are also pipelined between a single master–slave pair for both requests and responses. In principle, transactions can also be pipelined within a slave, if the slave allows this.

A transaction is composed of the following messages (see Fig. 4):

A *command* message (CMD) is sent by the ANIP, and describes the action to be executed at the slave connected to the PNIP. Examples of commands are read, write, test and set, and flush. Commands are the only messages that are compulsory in a transaction. For NIPs that allow only a single command with no parameters (e.g., fixed-size address-less write), we assume the command message still exists, even if it is implicit (i.e., not explicitly sent by the IP).

An *out data* message (OUTDATA) is sent by the ANIP following a command that requires data to be executed (e.g., write, multicast, and test-and-set).

A *return data* message (RETDATA) is sent by a PNIP as a consequence of a transaction execution that produces data (e.g., read, and test-and-set).

A *completion acknowledgment* message (RETSTAT) is an optional message that is returned by PNIP when a command has been completed. It may signal either a successful completion or an error. For transactions including both RETDATA and RETSTAT the two messages can be combined in a single message for efficiency. However, conceptually, they exist both to: RETSTAT to signal the presence of data or an error, and RETDATA to carry the data. In bus-based interfaces RETDATA and RETSTAT typically exist as two separate signals [1,12,13,17,18,26].

Messages composing a transaction are divided in *outgoing* messages, namely CMD and OUTDATA and *response* messages, namely RETDATA, RETSTAT. Within a transaction, CMD precedes all other messages, and RETDATA precedes RETSTAT if present. These rules apply both between master and ANIP, and PNIP and slave. Examples of transactions are shown in Fig. 5.

We classify connections as follows (see Fig. 6).

A *simple* connection is a connection between one ANIP and one PNIP.

A *narrowcast* connection is a connection between one ANIP and one or more PNIPs, in which each transaction that the ANIP initiates is executed by exactly one PNIP. An example of the narrow-cast connection is shown in Fig. 7, where the ANIP performs transactions

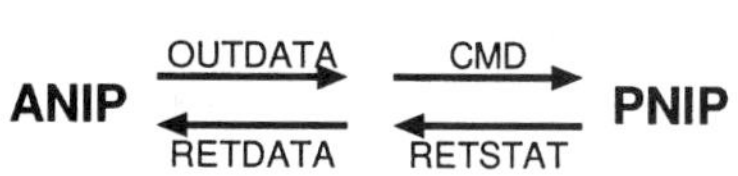

Figure 4 Transaction composition.

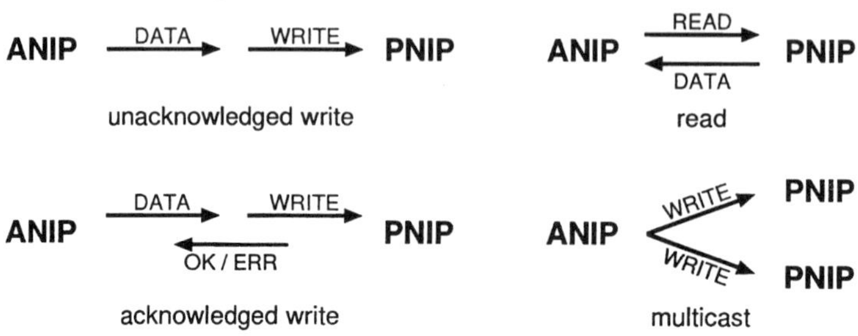

Figure 5 Transaction examples.

on an address space that is mapped on two memory modules. Depending on the transaction address, a transaction is executed on only one of these two memories.

A *multicast* connection is a connection between one ANIP and one or more PNIPs, in which the sent messages are duplicated and each PNIP receives a copy of those messages. In a multicast connection no return messages are currently allowed, because of the large traffic they generate (i.e., one response per destination). It could also increase the complexity in the ANIP because individual responses from PNIPs must be merged into a single response for the ANIP. This requires buffer space and/or additional computation for the merging itself.

B. Connection Properties

In this subsection we elaborate on the features that can be configured for a connection: guaranteed message integrity, guaranteed transaction completion, various transaction orderings, guaranteed throughput, bounded latency and jitter, and connection flow control.

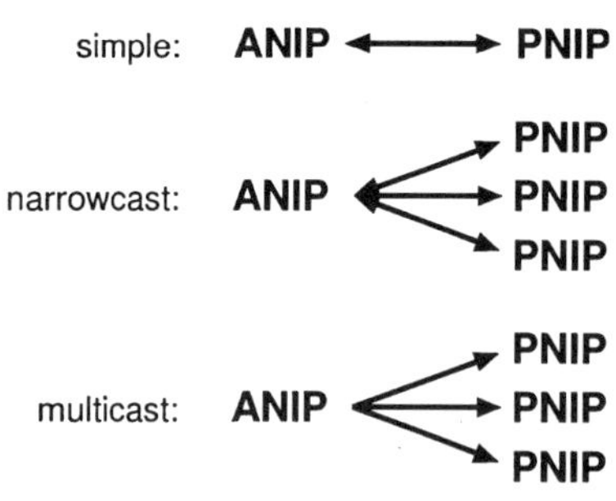

Figure 6 Connection types.

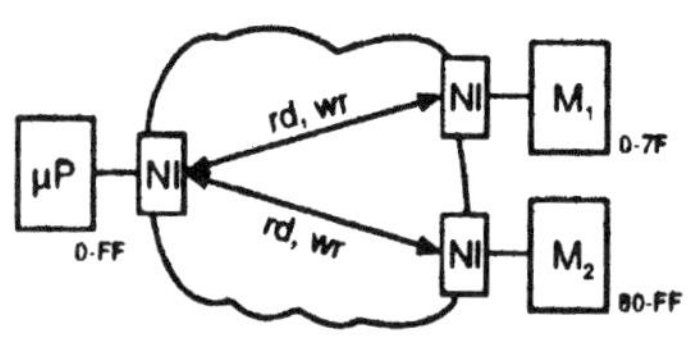

Figure 7 A narrowcast connection.

1. Data Integrity

Data integrity means that the content of messages is not changed (accidentally or not) during transport. We assume that data integrity is already solved at a lower layer in our network, namely at the link layer, because in current on-chip technologies data can be transported uncorrupted over links. Consequently, our network interface always guarantees that messages are delivered uncorrupted at the destination.

2. Transaction Completion

A transaction without a response (e.g., a posted write) is said to be complete when it has been executed by the slave. As there is no response message to the master, no guarantee regarding transaction completion can be given.

A transaction with a response (e.g., an acknowledged write) is said to be complete when a RETSTAT message is received from the ANIP*. The transaction may either (1) be executed successfully, in which case a success RETSTAT is returned, (2) fail in its execution at the slave, in which case an execution error RETSTAT is returned, or (3) fail because of buffer overflow in a connection with no flow control, in which case it reports an overflow error. We assume that when a slave accepts a CMD requesting a response, the slave always generates the response.

In our network, routers do not drop data [21]. Therefore, messages are always guaranteed to be delivered at the NI. For connections with flow control, NIs also do not drop data. Therefore, message delivery and, thus, transaction completion to the IPs is guaranteed automatically in this case.

However, if there is no flow control, messages may be dropped at the network interface in case of buffer overflow (see the paragraph on end-to-end flow control below). All of CMD, OUTDATA, and RETDATA may be dropped at the NI. To guarantee transaction completion, RETSTAT is not allowed to be dropped. Consequently, in the ANIPs enough buffer space must be provided to accommodate RETSTAT messages for all outstanding transactions. This is enforced by bounding the number of outstanding transactions.

3. Transaction Ordering

Across different connections no ordering of transactions is defined at the transport layer.

* Recall that when data is received as a response (RETDATA), a RETSTAT (possibly implicit) is also received to validate the data.

There are several points in a connection where order of transactions can be observed (see Fig. 8): (1) the order in which the master presents CMD messages to the ANIP, (2) the order in which the CMDs are delivered to the slave by the PNIP, (3) the order in which the slave presents the responses to the PNIPs, and (4) the order the responses are delivered to the master by the ANIP. Note that not all of (2), (3), and (4) are always present. Moreover, there are no assumptions about the order in which the slaves execute transactions; we can only observe the order of the responses. We consider the order of the transaction execution by the slaves to be a system decision and not a part of the interconnect protocol.

At both ANIP and PNIPs, outgoing messages belonging to different transactions on the same connection are allowed to be interleaved. For example, two write commands can be issued, and only afterwords their data. If the order of OUTDATA messages differs from the order of CMD messages, transaction identifiers must be introduced to associate OUTDATAs with their corresponding CMD.

Outgoing messages can be delivered by the PNIPs to the slaves (see Fig. 8b) as follows.

Unordered, which imposes no order on the delivery of the outgoing messages of different transactions at the PNIPs.

Ordered locally, where transactions must be delivered to each PNIP in the order they were sent (Fig. 8a), but no order is imposed across PNIPs. Locally ordered delivery of the outgoing messages can be provided either by an ordered data transportation, or by reordering outgoing messages at the PNIP.

Ordered globally, where transactions must be delivered in the order they were sent, across all PNIPs of the connection. Globally ordered delivery of the outgoing part of transactions require a costly synchronization mechanism.

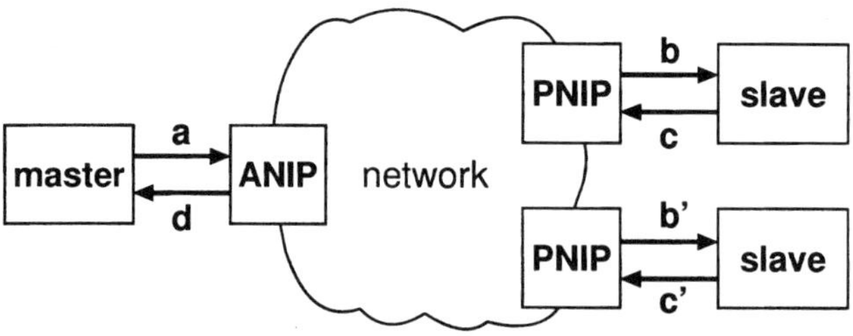

Figure 8 Message ordering is observable at a, b, c, and d.

Transaction response messages can be delivered by the slaves to the PNIPs (see Fig. 8c) as follows.

Ordered, when RETDATA and RETSTAT messages are returned in the same order as the CMDs were delivered to the slave (Figure 8b).
Unordered, otherwise.

When responses are unordered, there must be a mechanism to identify the transaction to which a response belongs. This is usually done using tags attached to messages for transaction identification (similar to tags in VCI).

Response messages can be delivered by the ANIP to the master (see Fig. 8d) as follows.

Unordered, which imposes no order on the delivery of responses. Here also, tags must be used to associate responses with their corresponding CMDs.
Ordered locally, where RETDATA and RETSTAT messages of transactions for a single slave are delivered in the order the original CMDs were presented by the master to the ANIP (Fig. 8a). Note that there is no ordering imposed for transactions to different slaves within the same connection.
Globally ordered, where all responses in a connection are delivered to the master in the same order as the original CMDs. When transactions are pipelined on a connection, then globally ordered delivery of responses require reordering at the ANIP.

All $3 \times 2 \times 3 = 18$ combinations between the above orderings are possible. Out of these, we define and offer the following two.

An *unordered* connection is a connection in which no ordering is assumed in any part of the transactions. As a result, the responses must be tagged to be able to identify to which transaction they belong. Implementing unordered connections has low cost, however, they may be harder to use, and introduce the overhead of tagging.

An *ordered* connection is defined as a connection with *local* ordering for the outgoing messages from PNIPs to slaves (Fig. 8b), *ordered* responses at the PNIPs (Fig. 8c), and *global* ordering for responses at the ANIP (Fig. 8d). We choose local ordering for the outgoing part because global ordering has a too high cost and few uses. The ordering of responses is selected to allow a simple programming model with no tagging. Global ordering at the ANIP is possible at a moderate cost, because all the reordering is done locally in the ANIP.

A user can emulate connections with global ordering of outgoing and return messages at the PNIPs, using nonpipelined acknowledged transactions, at the cost of high latency.

4. Connection Latency, Throughput, and Jitter

In our network, throughput can be reserved for connections in a time-division multiple access (TDMA) fashion, where bandwidth is split in fixed-size slots on a fixed timeframe. Bandwidth, as well as bounds on latency and jitter, can be guaranteed when slots are reserved. They are all defined in multiples of the slots. Throughput, latency and jitter can all be configured independently for the request and response parts of a connection.

Fully guaranteed-throughput connections (i.e., providing throughput guarantees on both request and return parts of the connection) can overbook resources in some cases. For example, when an ANIP opens a guaranteed-throughput read connection, it must reserve slots for the read command messages and for the read data messages. The ratio between the two can be very large (e.g., 1:100), which leads either to a large number of slots, or bandwidth, being wasted for the read command messages.

To resolve this problem, the request part of a connection can be best effort, while the response can have guaranteed throughput (or vice versa). For the example mentioned above, one can use best-effort read messages, and guaranteed-throughput read-data messages. No global connection guarantees can be offered in this case, but the overall throughput can be higher and more stable than in the case of using only best-effort traffic.

5. Connection Flow Control

As mentioned earlier, our network guarantees that messages are delivered to the NI. Messages sent from one of the NIPs are not immediately visible at the other NIP, because of the multi-hop nature of networks. Consequently, handshakes over a network would allow only a single message be transmitted at a time. This limits the throughput on a connection and adds latency to transactions. To overcome this problem, and achieve better network utilization, the messages must be pipelined. In this case, if the data is not consumed at the PNIP at the same rate it arrives, either flow control must be introduced to slow down the producer, or data may be lost because of limited buffer space at the consumer NI.

We introduce end-to-end flow control at the level of connections, which requires buffer space to be associated with connections. End-to-end flow control ensures that messages are sent over the network only when there is enough space in the NIP's destination buffer to accommodate them.

End-to-end flow is optional (i.e., to be requested when connections are opened) and can be configured independently for the outgoing and return paths. When no flow control is provided, messages are dropped when buffers overflow. Multiple policies of dropping messages are possible, as in off-chip networks. Possible scenarios include the oldest message is dropped (milk policy), or the newest message is dropped (wine policy) [25].

We opt for a credit-based flow control. Credits are associated with the empty buffer space at the receiver NI. The sender's credit is lowered as data is sent. When the PNIP delivers data to the slave, credits are granted to the sender. If the sender's credit is not sufficient to send some data, the NI at the sender stalls the sending.

C. Use Cases

To illustrate the need for differentiated services on connections, we consider two examples of traffic. We describe the properties they would use over an Æthereal connection to meet their traffic requirements.

Video processing streams typically require a lossless, in-order video stream with guaranteed throughput, but possibly allow corrupted samples. An Æthereal connection for such a stream would require the necessary throughput, ordered transactions, and flow control. If the video stream is produced by the master, only write transactions are necessary. In this case, with a flow-controlled connection there is no need to also require transaction completion, because messages are never dropped, and the write command and its data are always delivered at the destination. Data integrity is always provided by our network, even though it may not be necessary in this case.

Another example is that of cache updates which require uncorrupted, lossless, low-latency data transfer, but ordering and guaranteed throughput are less important. In this case, a connection would not require any time related guarantees, because even though an average low latency is required, a guarantee on low latency is not critical. Low latency can be obtained even with a best-effort connection. The connection would also require flow control and guaranteed transaction completion to ensure lossless transactions. However, no ordering is necessary, because this is not important for cache updates, and allowing out-of-order transactions can reduce the response time.

V. CONCLUSIONS

In this chapter, we compared networks on chip (NoC) to off-chip networks (e.g., computer networks) and existing on-chip interconnects (e.g., buses). We showed that NoCs have many similarities with off-chip networks. However, they also differ, especially in their resource constraints. For example on a chip, memory and computation resources are more expensive, while there are more wires. This makes NoC architectures different from off-chip networks, and requires rethinking of network services.

We also compared NoCs to existing on-chip interconnects, such as buses and switches. By directly connecting IP blocks, existing on-chip interconnects can offer tight coupling between masters and slaves, and global arbitration. In NoCs, masters and slaves are completely decoupled, and the arbitration is distributed over the network nodes. This makes it harder to provide guarantees, such as bandwidth lower bounds, and transaction orderings.

We defined a set of NoC services that abstract from the network details. Using these services in IP design decoupled computation and communication. We used a request–response transaction model to be close to existing on-chip interconnect protocols. This eases the migration of current IPS to NoCs. To fully utilize the NoC capabilities, such as high bandwidth and transaction concurrency, our services provide connection-oriented communication. Connections can be configured independently with different properties. These properties include transaction completion, various transaction orderings, bandwidth lower bounds, latency and jitter upper bounds, and flow control.

Our NoC services are a prerequisite for service-based system design, which makes applications independent of NoC implementations, makes designs more robust, and enables architecture-independent quality-of-service strategies.

REFERENCES

1. ARM (1999). *AMBA Specification. Rev. 2.0.*
2. ARM (2001). *Multi-Layer AHB. Overview.*
3. Bainbridge, J., Furber, S. (2002). CHAIN: A delay-insensitive chip area interconnect. *IEEE Micro* 22(5).
4. Benini, L., De Micheli, G. (2001). Powering networks on chips. In: *ISSS.*
5. Benini, L., De Micheli G. (2002). Networks on chips: A new SoC paradigm. *IEEE Computer* 35(1):70–80.
6. Culler, D. J., Singh, J. P., Gupta, A. (1999). *Parallel Computer Architecture: A Hardware/Software Approach.* Morgan Kaufmann Publishers.
7. Dally, W. J., Towles, B. (2001). Route packets, not wires: On-chip interconnection networks. In: *DAC.*
8. Goossens, K., van Meerbergen, J., Peeters, A., Wielage, P. (2002). Networks on silicon: Combining best-effort and guaranteed services. In: *DATE.*
9. Guerrier, P., Greiner, A. (2000). A generic architecture for on-chip packet-switched interconnections. In: *DATE.*
10. Having, P.J. (2000). *Mobile Multimedia Systems.* PhD thesis, University of Twente.
11. Hennessy, J., Patterson, D. (1995). *Computer Architecture: A Quantitative Approach.* Morgan Kaufmann.
12. IBM (2001). *32-bit On-Chip Peripheral Bus.* Rev. 2.I.
13. IBM (2001). *32-bit Processor Local Bus.* Rev. 2.9.
14. Keutzer, K., Malik, S., Newton, A. R., Rabaey, J. M., Sangiovanni-Vincentelli,

A. (2000). System-level design: Orthogonalization of concerns and platform-based design. *IEEE Trans. on CAD of Integrated Circuits and Systems* 19(12):1523–1543.
15. Kumar, S., Jantsch, A., Soininen, J.-P., Forsell, M., Millberg, M., Öberg, J., Tiensyrjä, Hemani, A. (2002). A network on chip architecture and design methodology. In: *ISVLSI*.
16. Leijten, J. A., van Meerbergen, J. L., Timmer, A. H., Jess, J. A. (1997). Prophid, a data-driven multi-processor architecture for high-performance DSP. In: *ED&TC*.
17. OCP International Partnership (2001). *Open Core Protocol Specification.*
18. Philips (2001). *DTL Protocol Specification.* Rev. 2.I.
19. Rexford, J. (1999). *Tailoring Router Architectures to Performance Requirements in Cut-Through Networks.* PhD thesis, Univ. Michigan.
20. Rijpkema, E., Goossens, K., Rădulescu, A., Dielissen, J., van Meerbergen, J., Wielage, P., Waterlander, E. (2003). Trade offs in the design of a router with both guaranteed and best-effort services for networks on chip. *DATE.*
21. Rijpkema, E., Goossens, K., Wielage, P. (2001). A router architecture for networks on silicon. In: *PROGRESS.*
22. Rose, M. T. (1990). *The Open Book: A Practical Perspective on OSI.* Prentice Hall.
23. Sgori, M., Sheets, M., Keutzer, K., Malik, S., Rabaey, J., Sangiovanni-Vincentelli, A. (2001). Addressing the system-on-a chip interconnect woes through communication-based design. In: *DAC.*
24. Stravers, P., Hoogerbrugge, J. (2001). Homogeneous multiprocessing and the future of silicon design paradigms. In: *VLSI-TSA.*
25. Tanenbaum, A. S. (1996). *Computer Networks.* Prentice Hall.
26. VSI Alliance (2000). *Virtual Component Interface Standard.*
27. Zhang, H. (1995). Service disciplines for guaranteed performance service in packet-switching networks. *Proc. of the IEEE* 83(10):1374–1396.

11

Single-Chip Multiprocessing for Consumer Electronics

Paul Stravers and Jan Hoogerbugge
Philips Research Laboratories, Eindhoven, The Netherlands

I. INTRODUCTION

The consumer and telecom semiconductor industry is facing two challenges that need immediate attention. The first challenge is to keep up with the ever-increasing pace of product launches. The globalized economy increases the pressure of competitors while bored consumers have acquired a taste for action and sensation that needs regular reinforcement with new gadgets. Because the markets are unpredictable, a large share of product launches will turn out as failures, but the small fraction that does turn into a success can be sold in high volume and with high margin.

The second challenge for the semiconductor industry is to control the ever-increasing design complexity of chips, where Moore's law rules, dictating transistor counts running into the hundreds of millions. These engineering projects must take on a wide range of problems, including failing design automation tools and strategies, unqualified engineers, insufficient circuit simulation resources, and serious issues with project quality and management.

The aggregate of these challenges raises a very serious issue for the industry, namely how to deliver these extremely complex products to a fast-moving, competitive, and globalized market. In fact it seems plausible that only a handful of semiconductor companies will be able to successfully tackle this problem. The ones that do not will be confined to niche markets and low-margin products.

II. MOTIVATION

We propose to take on the challenges with a chip architecture that trivially scales with Moore's law due to its regular, homogeneous morphology. In the past, similar proposals failed or were confined to niche markets, mainly because they lacked computational efficiency in a sufficiently wide range of applications. To avoid this trap, we relied on a technology trend that relates the amount of silicon area spent on memory to silicon area spent on computing logic. This trend can be exploited to achieve high computational efficiency throughout the design hierarchy, while preserving the homogeneous regularity at the top level of chip architecture.

Embedded computer chips exhibit a trend where with every new generation an increasing percentage of the chip area is dedicated to memory, while an ever-decreasing percentage of the chip area is dedicated to computational structures. This observation can be rationalized as follows. It has long been known that a balanced computer system is equipped with an amount of memory that is proportional to the computational power of the processing unit [1]. Richard Case observed that mainframe computers follow the rule of 1 memory byte per instruction per second. Like Moore's law, Case's ratio has no rigorous foundation but it has held remarkably well over the four decades since it was first postulated.

To see how Case's ratio affects the ratio of computational resources to memory resources on a chip, we note that each new generation of semiconductor process technology reduces the area of both computational and memory structures by a factor A, while increasing the maximum achievable clock frequency of a chip by a factor S.

We introduce ρ, the ratio of memory area M to compute area C. For the left-hand side of Fig. 1 this number is

$$\rho_{t_0} = \frac{M_{t_0}}{C_{t_0}} \tag{1}$$

After some time $\tau = t_1 - t_0$ a new process generation is introduced and we scale both C_{t_0} and M_{t_0} by a factor A. We maintain the property of a balanced computing system by matching the now S times faster compute logic to an S times larger memory. The result is depicted on the right-hand side of Fig. 1 where we find

$$\rho_{t_1} = \frac{S(M_{t_0}/A)}{C_{t_0}/A} = S\rho_{t_0} \tag{2}$$

Assuming that new technology generations come at more-or-less regular intervals τ, as was the case with CMOS-technology up to now, we find

$$\rho_t = \rho_0 S^{\rho t/\tau} \tag{3}$$

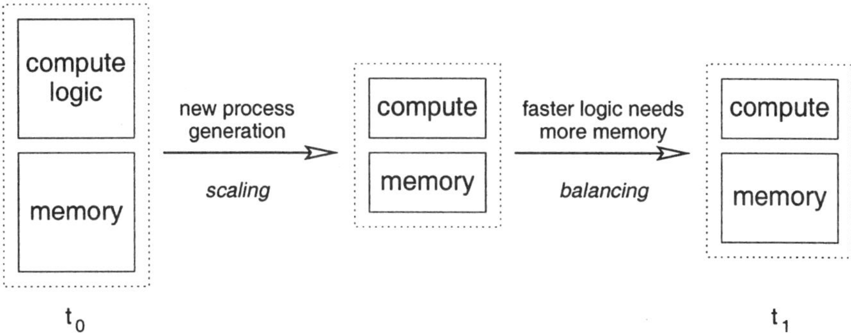

Figure 1 Effect of technology scaling on the memory-to-compute ratio.

Note that independent from its initial value, ρ_t increases exponentially over time as long as new silicon process generations are introduced with $S > 1$. For CMOS-technology in the past decade, the value of S has been close to 1.4 and the value of τ has been close to 2 years.

Eq. 3 points to the conclusion that in the future memory will increasingly dominate the available silicon area, while compute logic will evolve into a small fraction of the available area. Because the compute logic is getting so small and the memories so big, the average wiring distance between the two is becoming relatively large, resulting in reduced system performance and increased power consumption. This problem can be addressed by clustering the chip space, where each cluster contains a share of the compute logic connected to a share of the memory. The clusters are small enough that their electrical properties are not dominated by the effect of long wires.

An equally important consequence of a high memory-to-compute ratio is the observation that compute logic is very cheap in terms of area and therefore we can afford to include compute logic on the chip and *then not use it*. This means that a cluster can be equipped with a large number of special purpose hardware functions that achieve high computational efficiency in different but narrow application domains.

For example, a cluster could be equipped with hardware functions that perform video stream processing, plus hardware to perform telecom filtering operations, plus hardware to perform graphics manipulations, plus some general purpose DSP and microcontrollers, and so on. Clusters have the ability to enable or disable each hardware function *after* manufacturing. Later, when an application is mapped onto the chip it may turn out that the memory consumption of the application's algorithm only allows the use of the telecom filters in a particular cluster, plus one DSP and nothing else. In this case all other hardware functions in the cluster simply remain unused.

Because of the high memory-to-compute ratio, wasting most of the compute resources this way is not very expensive. On the other hand, the most valuable resource in a cluster is memory, so the application must be distributed over the clusters in a way that maximizes the memory utilization.

The advantage of such an organization is that we have field programmable hardware without the usual drawback of low computational efficiency. The computational efficiency is not quite as high as it would have been in a dedicated solution because we have introduced the overhead of a configurable switch that is needed to connect the subset of selected hardware functions to the memory and to each other.

Finally, because every cluster on the chip is capable of efficiently executing a wide range of algorithms, there is no good reason why one cluster would need to be equipped with a different set of hardware functions than any other cluster. If all clusters are designed equal we arrive at the homogeneous morphology that is so desirable for trivial design scaling when new technology generations are introduced.

An additional but important benefit of a homogeneous architecture is redundancy. If a certain chip was designed with n clusters, then if one cluster is nonfunctional because of fabrication faults, we still have a chip with the same functionality as the fault-free chip. The only difference is that the chip with defects has a lower computational power. This need not be a problem because we only match the chip with an application after fabrication. The faulty chips are only used for applications that do not demand the highest available performance.

From an economic point of view this redundancy means that faulty chips still have a high economic value. The *yield* of a large homogeneous multiprocessor can be close to 100%, even if traditional yield models would have predicted a low yield for chips of that same large size.

III. ARCHITECTURE

The observations made in the previous section suggest a regular structure of communicating *tiles* (the uniform clusters). Each tile can be configured to execute a set of *tasks*. Assignment of tasks to tiles is statically determined, but within a tile tasks can dynamically arbitrate for resources such as memory and the special purpose coprocessors. Section IV discusses the programming paradigm from which the tasks are derived.

The details of a suitable inter-tile communication infrastructure is one of our main research topics. In our CAKE project (Computer Architecture for a Killer Experience, referring to the sensation that CAKE applications induce in the user's mind) we mostly concentrated on a two-dimensional torus, see Fig. 2.

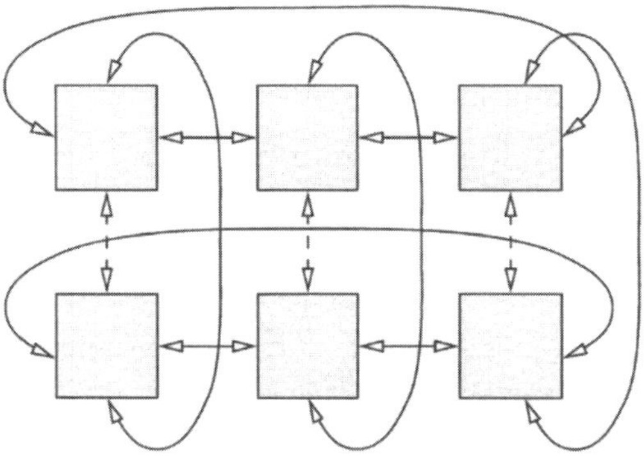

Figure 2 Homogeneous network of tiles.

As discussed above, the size of the tiles should be small enough so that they do not suffer too much from long intra-tile wiring. But the tiles should be large enough to host a significant number of hardware functions to achieve high levels of computational efficiency on a wide range of applications.

Figure 3 depicts a typical tile design. The blocks labeled SPF represent the special purpose hardware functions that are key to the computational efficiency of CAKE chips. Section V discusses the interaction of the SPF coprocessors and software tasks running on the CPU. There are multiple memory banks to increase the concurrency and improve throughput. All

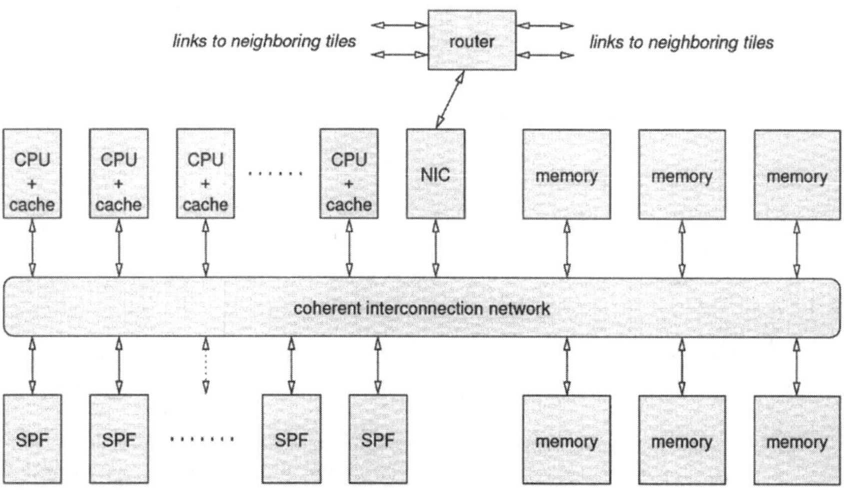

Figure 3 Typical architecture of a tile.

communication with other tiles on the chip is done by the router. The NIC is the network interface controller, responsible for the OSI *network* and *transport* layers in the communication protocol.

Although CAKE is agnostic with respect to CPU instruction set architectures, it does require that the CPU support the same bus standard and the same cache coherence protocol. The CAKE instances that we studied include a mix of low-power MIPS, mid-range MIPS, and high-performance Trimedia [2] CPU. Both MIPS and Trimedia support the same data types and data alignment (e.g. 32-bit integer, 8-bit integer, etc.), which makes it easy to communicate across the various instruction set domains without the overhead of an architecture-independent data interchange format. For example, it is perfectly all right to share a semaphore between a MIPS task and a Trimedia task, and therefore a state change in the MIPS domain can unblock a process in the Trimedia domain with no need for pesky interrupts.

IV. PROGRAMMING

Today's compiler technology is far from being capable to map sequential application code written in C or C++ efficiently on a highly parallel architecture as the CAKE architecture. Therefore, parallelism must be expressed explicitly in the application code.

CAKE supports various parallel programming paradigms. The most popular are POSIX threads (also known as pthreads) and YAPI process networks [3]. Pthreads is an industry standard for parallel programs but it is only useful for software applications. Process networks have not evolved yet into an industry standard, but they allow the specification of mixed hardware–software systems.

For complex consumer applications CAKE supports mixed programming paradigms: at the highest hierarchical level the application expresses task level parallelism using YAPI process networks. Going down in the hierarchy each YAPI process can contain a complete YAPI subgraph, or it can contain multiple POSIX threads, or it can contain a simple sequentially executing process.

In this chapter we concentrate on YAPI process networks. A YAPI case study is presented in Section VI. A case study with pthreads is presented in Section VII.

A. Process Networks

YAPI process networks are a derivative of Kahn process networks [4]. A Kahn process network is a directed graph where nodes correspond to sequential processes and edges correspond to FIFOs. All sequential processes run in parallel and only communicate and synchronize with each other by reading from or writing to FIFOs. Processes block when reading from an empty FIFO.

In the original Kahn definition, FIFOs are infinitely large, so processes can never block a write operation. Since this is a little impractical for an embedded system with limited resources, YAPI limits the FIFO size and consequently introduces block-on-write. Unfortunately this also introduces the possibility of artificial deadlock. This is a state where the process network does not make progress, not because of a true deadlock but simply because a critical process is blocked while attempting to write to a full FIFO. CAKE remedies this situation by detecting artificial deadlocks at run-time and then enlarging the critical FIFO [5].

The process network programming paradigm is a simple and convenient method to code signal processing applications. A signal processing application is decomposed in well-defined computations. These computations become processes that read their operands from incoming FIFOs, compute results, write these results to outgoing FIFOs, and typically repeat this in an endless loop.

Because interaction with the environment is simple and well-defined, the code that implements these computations lends itself for reuse in other applications. Many semiconductor companies have attempted to introduce the concept of core reuse, where dedicated hardware cores can be reused in many projects. However, in practice this goal is only achieved as long as only a few reusable cores are integrated in a system; typically a reusable core comes with a fat user manual and complex rules for interaction (e.g., a device driver is needed and interrupt lines and DMA channels, etc.). This complexity makes it very hard to build systems with more than several tens of such "reusable" cores. Yet the technology curve continues to provide ever-larger numbers of transistors on a chip and, therefore, the current approach to core reuse is doomed. The properties of process networks make them much better suited for reuse than VLSI cores.

Note that Kahn process networks are different from dynamic data flow networks. In particular, an actor in a DDF network has a more explicit input/output behavior when compared to a Kahn process. Although in general DDF networks are more expressive than Kahn networks, their drawback is the need for explicit implementation choices in the design of the actors. For example, actors that are more powerful than Kahn processes typically need to maintain an explicit internal state machine. Kahn processes on the other hand contain fewer of such implementation choices and therefore can be considered more abstract and therefore more suitable to behavioral specifications.

In the past the abstract nature of Kahn processes has been problematic because it often leads to inefficient implementations. The CAKE architecture and its run-time system have been carefully tuned to efficiently perform process context switches and to support an abundance of low-cost, high-speed semaphore operations, all of which are critical to the efficient execution of Kahn process networks.

To illustrate this, it only takes 96 cycles from the moment that a process starts a `write ()` operation to the moment that the next process starts running on that same processor. In those 96 cycles, semaphore operations, state saving, schedule activity and state loading take place.

Coding YAPI applications for CAKE is done through a library of C++ classes. For example, a new process `foo` is created by deriving `foo` from the YAPI `Process` base class. Besides being available for the CAKE architecture, YAPI implementations are also available for workstations so that applications can be developed and tested on workstations before being ported to CAKE.

B. Implementation

Mapping of CAKE applications consists of mapping processes on processors. Objectives are load balancing and minimizing communication. Mapping of CAKE applications is a two-stage activity: first processes are mapped on tiles, and second, processes within tiles must be mapped on processors. Tile allocation is performed statically, while processor allocation within a tile is performed dynamically. The motivation for this is that dynamic tile allocation is considered to be too complex in the absence of a high-bandwidth shared memory between the tiles. Furthermore, dynamic processor allocation is feasible and typically gives better load balancing than static processor allocation.

In order to perform a good tile allocation, the CAKE application is profiled and the computational load of every process and the communication load through every FIFO is measured. The result is a process network with weights assigned to processes and FIFO. This is the input of the tile allocator together with a description of the target CAKE architecture. The tile allocator solves a combinatorial optimization problem where the execution time, determined by the heaviest loaded tile or inter-tile link, is minimized.

Processor allocation, or process scheduling, is performed dynamically. Every tile has a pool of ready processes, of which the processors repeatedly take a process and execute it until it blocks on a FIFO operation. Whenever the process is unblocked, it may be rescheduled on another processor. Measures are taken to prevent that processes move too frequently between processors, which would result in excessive cache coherence traffic between the processors.

Process scheduling and FIFO operations are implemented very efficiently. The efficiency determines how fine-grain processes can communicate with each other. In the CAKE implementation it only takes 97 RISC instructions to resume a runable process after another process blocked on a FIFO operation.

FIFOs consist of buffer space, a read pointer, a write pointer, a semaphore `data` corresponding to the number of valid elements in the buffer, and a semaphore `room` corresponding to the number of free spaces in the buffer. The semaphores are used to implement blocking on reading from an empty FIFO and writing to a full FIFO. Read and write operations on an intra-tile FIFO operate directly on the FIFO as shown below.

```
template<class T> void FIFO: :write(T value){
    P(room); buffer[ wptr++] = value; V(data);
}

template<class T> void FIFO: :read(T &value){
    P(data); value = buffer[rptr++] ; V(room);
}
```

In the case of inter-tile FIFOs, the `room` semaphore is stored at the writer-side, while the other FIFO data items are stored at the reader-side. A write operation performs a P operation on `room` before it sends an active message [6] containing the data to be written to the reader-side. The active message handler stores the data in the buffer space and performs a V operation on `data`. A read operation performs a P operation on `data` reads data from the buffer space and sends an active message to the writer-side. This active message performs a V operation on `room`. These active messages back to the writer-side implement flow-control between writer and reader. Active messages of successive reads can be combined in order to reduce communication.

V. COPROCESSOR CONTROL

As described in the introduction, processes are either running on a programmable processor or they run on a dedicated coprocessor. Which option to chose for every process is the classical hardware/software codesign question. In designing the FIFO communication protocol we also asked the question of what to do in hardware and what in software in order to obtain a flexible, powerful, cost-efficient, and easy-to-design system. We wanted to do things in software that is complex to do in hardware or when there is no real need to do it in hardware. Applying this to FIFO communication leads to the following division: buffer access, pointer increments, and pointer wrap-around tests in hardware and semaphore operations in software. It is desirable to perform semaphore operations in software because of their complexity (waking up sleeping processes for example).

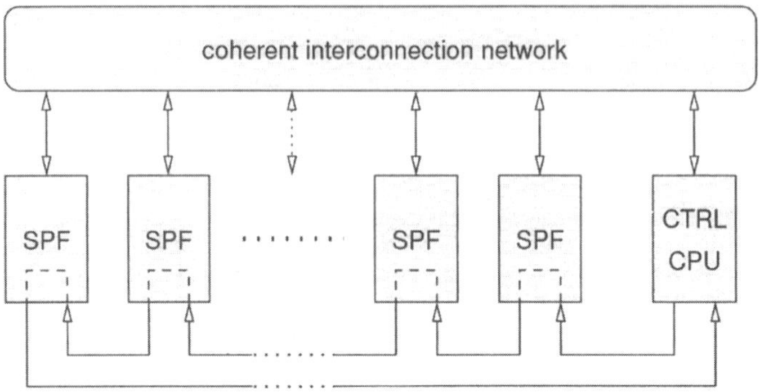

Figure 4 The coprocessor control infrastructure.

Figure 4 focuses on the SPF (coprocessor) part of CAKE. This consists of a multitude of special purpose hardware (labeled SPF in the figure), plus a controller CPU. The controller CPU performs semaphore operations on behalf of the coprocessors. To this end control messages are exchanged between the controller CPU and the coprocessors by means of a dedicated ring network.

The pseudo code for writing data to a FIFO on a coprocessor looks as follows (note that this pseudo code is implemented as part of the coprocessor hardware).

```
void write (fifo, val)
{
    while (room_available = 0)
        room_available = get_room (fifo);

    store (wptr++, val);
    if (wptr == buffer_end)
        wptr = buffer_begin;

    room_available-;
    data_produced++;

    if (data_produced == data_produced_limit)
    {
        put_data (fifo, data_produced);
        data_produced = 0;
    }
}
```

To avoid that a coprocessor must request the controller for two semaphore operations (a P and a V) on every read or write, we maintain two counters for a write port to a FIFO. First, `room_available` contains the number of free spaces that we can write to before we ask the controller for free space. Second, `data_produced` contains the number of data values that we have written to the FIFO buffer but have not been reported to the controller yet.

Now back to the pseudo code for writing to a FIFO from a coprocessor above. First we check the `room_available` counter. If it is zero we ask the controller for room. Because the controller may return that no room is available, we keep asking until `get_room` returns non-zero. Next, we store the data in memory, increment the write pointer and test for wrap around. After decrementing `room_available` and incrementing `data_produced` counters, we test whether enough data has been produced to inform this to the controller by means of a `put_data` message. The `put_data` message gets as argument how many data have been produced.

The pseudo code for reading from a FIFO is similar to the code for writing. Instead of `get_room` and `put_data` messages, the coprocessor sends `get_data` and `put_room` messages to the controller.

Now the software that runs on the controller. The controller has to accept four types of messages: `get_room`, `get_data`, `put_room`, and `put_data`. `get_room` is implemented as follows.

```
int get_room(struct FIFO *fifo)
{
    return sem_reset(&fifo->room);
}
```

`get_room` calls `sem_reset` with the room semaphore of the FIFO as argument. `sem_reset` resets the value of the semaphore to zero and returns the previous value of the semaphore. This value is returned to the coprocessor that sent the `get_room` message.

`put_data` is implemented as follows. It increments the data semaphore of the FIFO by the specified value. Vn(s, n) is semantically equivalent to doing *n* times V(s).

```
void put_data(struct FIFO *fifo, int count)
{
    Vn(&fifo->data, count);
}
```

The implementation of `get_data` and `put_room` are similar to `get_room` and `put_data` respectively.

The CAKE coprocessors improve on the described communication scheme by requesting data/room to the coprocessor before the `data_available` / `room_available` have become zero. This hides the latency of the `get_data` / `get_room` message.

VI. CASE STUDY: PROCESS NETWORKS

An important aspect of application development for the CAKE architecture is the design of process networks. Typically an application can be expressed as many different process networks, all of which are functionally equivalent but with varying degrees of exposed parallelism. In this section we describe experiences in the development of a YAPI MPEG2 decoder for CAKE.

One of the most important lessons that we learned is to avoid tight cycles in the process network graph. An extreme example is shown in Fig. 5. If process A is sending commands to process B, process B is sending results back to A, and A is waiting while B executes the commands, then the parallelism is effectively less than one (i.e., less than one process will be running at a time). One way to improve this situation is to allow for multiple outstanding commands. An alternative solution is to merge processes A and B.

Another important aspect of mapping applications on the CAKE architecture is to introduce data parallelism next to task parallelism. Consider the top process network fragment of Fig. 6. If the IDCT process turns out to be a performance bottleneck, we can create multiple IDCT processes each handling a share of the data to be processed. In this example each IDCT process has its own input and output FIFO and the IQ and ADD processes write to and read from three FIFOs in a round robin fashion. Alternative solutions with shared FIFOs with multiple readers and/or writers are also possible.

Figure 7 shows the performance of the MPEG2 decoder after applying the transformations discussed above. The speedup of an 8-CPU configuration is 5.4. Moving to a 16-CPU configuration does not improve the speedup much and should therefore be considered a waste of resources.

Figure 7 shows the phenomenon that speedup benefits more from an increased number of CPUs per tile than it benefits from an increased number of tiles. For example, the speedup of a "4 tiles $\times$ 1 CPU" configuration is 2.6, compared to 3.5 for a "1 tile $\times$ 4 CPUs" configuration. At first it seems that this is caused by the fact that the "4 $\times$ 1" configuration only uses the message passing network to communicate among the CPUs, while the "1 $\times$ 4" configuration only uses shared memory for communication.

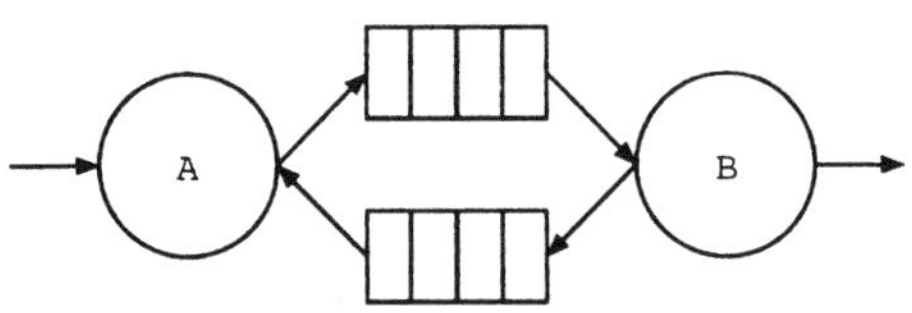

Figure 5 A tight cycle in a process network.

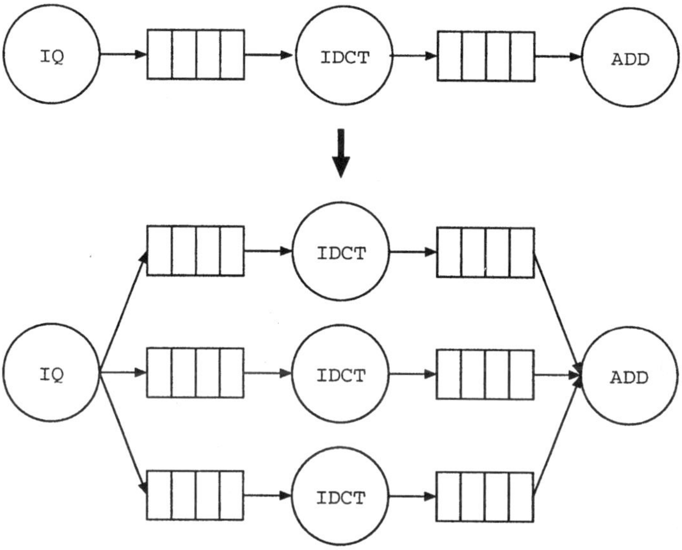

Figure 6 Introducing data parallelism.

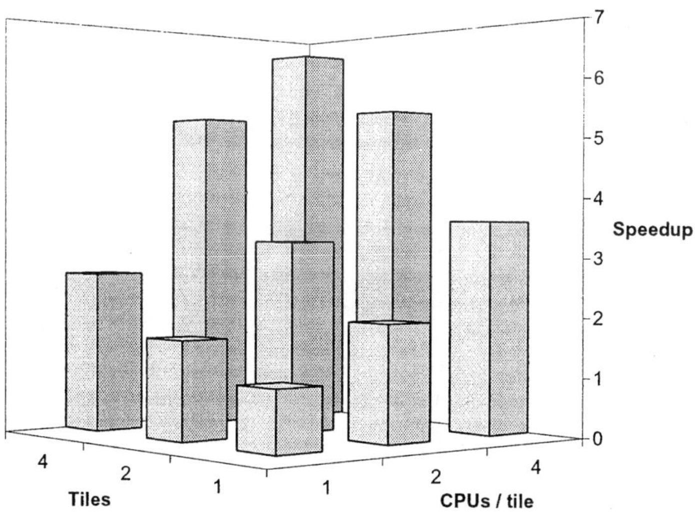

Figure 7 Speedup of the YAPI MPEG2 application.

However, we found that the difference in speedup can mainly be attributed to the different scheduling opportunities in the two configurations. The static allocation of YAPI tasks to a tile prevents that a ready-to-run task on tile A can execute on an idle processor of tile B. As a consequence the "4 × 1" configuration experiences idle CPU cycles even at times when ready-to-run tasks are waiting on tiles with a busy CPU. In contrast, the dynamic scheduler running on the "1 × 4" configuration can keep all 4 CPUs busy as long as there are at least 4 ready-to-run tasks in the YAPI application graph.

VII. CASE STUDY: MULTITHREADING

The case study in the previous section exploits simple task-level parallelism by pipelining various MPEG2 functions (e.g., we pipelined IQ, IDCT and ADD). But it appears that this direction is not very promising because the degree of pipelining (and hence the potential for parallelism) is not high (i.e., at most 3 or 4). With additional data parallelism as shown in Fig. 6 the degree of parallelism improved to 6 or 7.

In this second case study we focused almost completely on data parallelism for high definition (HD) video decoding. To reduce synchronization overhead we wanted to increase the granularity of the parallel tasks. In addition, we studied how difficult it is to express data parallelism with industry standard POSIX threads. Supporting industry standard programming interfaces is important because it opens up the CAKE architecture to a large installed base of applications.

We downloaded the original nonthreaded MPEG2 decoder from the `mpeg.org` website. We then studied the MPEG2 algorithm, looking for opportunities to introduce parallelism.

A. Code Changes for Multi-Threading

Each 1920 × 1088 MPEG2 high definition picture is composed of at least 68 *slices* [7]. Intended as a provision for easy transmission error recovery, each slice can be processed independently from all other slices in the same picture. Moreover, the start of a new slice in the picture data is announced by a unique bit pattern called the *start code*.

So our strategy for a multi-threaded MPEG2 decoder is the following. The main thread handles all the higher layers of the MPEG2 stream, down to the picture data. The original function to handle the picture data looks as follows.

```
while(next_start_code())    /* search for unique start
code*/
    slice ();
```

Instead of handling slices sequentially, we wanted to handle them concurrently, processing each slice in its own thread.

```
while (next_start_code()){
    context[i] = copy_context ();
    create_thread(slice, &context[i]);
    ++i;
}
wait_for_all_unfinished_threads();
```

The `slice ()` function refers to global state, and sometimes it even updates global state. For example, while scanning through the compressed video input buffer, it updates the "current position" pointer. Clearly, with multiple `slice ()` threads running concurrently there are multiple "current" positions at the same time. Therefore each thread maintains its own context, a struct with all state variables that must be kept local to the thread. The `slice ()` context is initialized by the main thread simply by copying its own context as it appears when it detects a new slice header in the video input stream. This is the function of `copy_context ()`. The context structs exist in the name space of the main thread, and they are recycled after a `slice ()` thread dies.

Some references to previously global variables in `slice ()` must now be converted to references in the thread context. The hardest part really is finding out what globals are potentially modified by `slice ()`. In our experiment we did this by reasoning about the MPEG2 algorithm, but there are also tools in development at Philips Research, Eindhoven to perform such analysis automatically.

In the case of the global input buffer, it requires nontrivial intervention to make it thread-safe. The problem with the original code is that it has only one buffer and just after the last byte is consumed the buffer is refilled with the next block of input data. This is not acceptable because it is very likely that there are still other threads with a "current position" pointer that refer to the old data.

Our solution to this problem is simple: we replace the existing input buffer with a linked list of buffer segments. Only the main thread can allocate and fill new segments; the `slice ()` threads only follow the links. The main thread only recycles a buffer segment after it is sure that all threads referring to that segment have died. The code for the new buffer scheme is less than 75 lines.

B. Measurements

We run the multi-threaded application on a single CAKE tile with a variable number of Trimedia [2] processors.

The success of our effort is measured with two metrics. The first is speedup as a function of the number of Trimedia CPUs (see Fig. 8). The second metric is the penalty incurred for making the original code thread-safe. This is measured as the quotient of the execution times of the thread-safe code and the original code, both running on a single Trimedia. This number is 1.08.

Figure 8 shows an almost linear speedup for multiprocessors with up to 9 Trimedias. Beyond that point the graph quickly levels off, and in fact it never exceeds a speedup larger than 10. In the next section we analyze the cause of this behavior and we suggest a simple remedy.

Figures 9 and 10 show thread activity during the first three pictures of the HD stream. They clearly show the `slice ()` threads running concurrently during picture data. The concurrency disappears completely between subsequent pictures, as shown by the steep V-shapes at 1/3 and 2/3 of the time line. Note that the main thread is not shown in the figures, which explains why only 7 threads are active with 8 CPUs. Toward the end of each picture the main thread is blocked (waiting for threads to finish) so then all 8 CPUs can run `slice ()` threads.

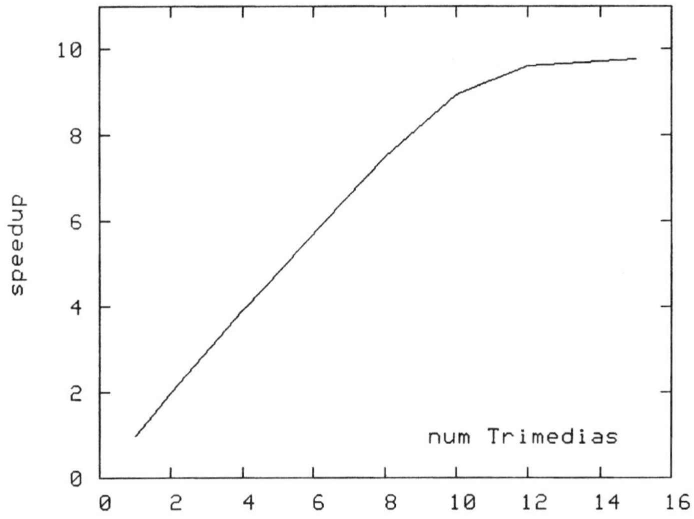

Figure 8 Speedup with no optimizations.

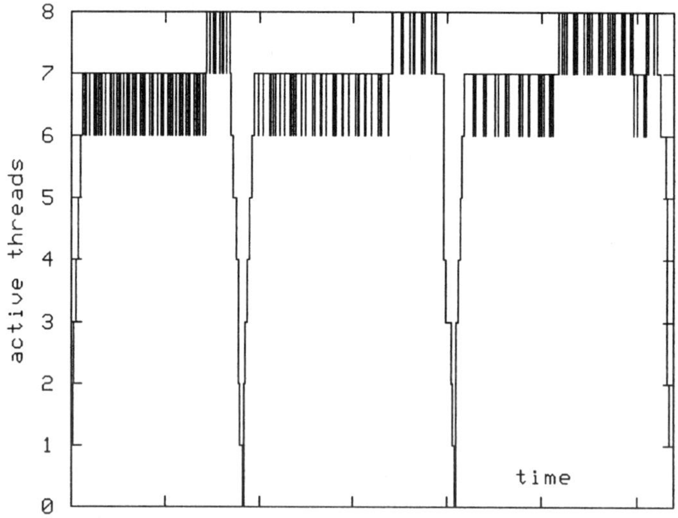

Figure 9 Thread concurrency with 8 CPUs.

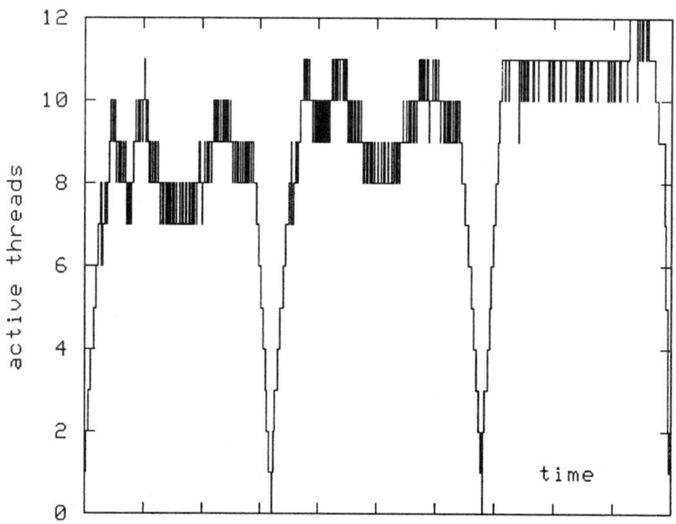

Figure 10 Thread concurrency with 12 CPUs.

C. Interpretation

Beyond 10-way multiprocessing the main thread becomes the bottleneck. It cannot find new slices fast enough to keep all CPUs busy all the time. The critical loop in the main thread looks like this (with CPU cycle consumption in comments).

```
while (next_start_code()) {                         /* 200k */
    context[i] = copy_context();                    /*   4k */
    create_thread(slice, &context[i]);              /*   1k */
    ++i;
}
```

Clearly the bottleneck is a very inefficient implementation of `next_start_code ()`: it consumes 45 cycles to scan a single byte from the compressed video input buffer! With a little bit of care it is possible to reduce this number to only 2 or 3 cycles per byte (20x improvement!).

When `next_start_code ()` is fixed, the next thing to look at is `copy_context ()`. A close examination shows that in fact 90% of the context struct that we copy needs no initialization at all, and so it is possible to reduce this function to less than 500 cycles.

As a consequence, the optimized main loop is able to start a new thread every 12.5k cycles (11k + 0.5k + 1k), a 16x improvement. In that case the main loop can keep up with a real-time HD video stream using only 26 MHz of a Trimedia CPU (coming from 400 MHz in the unoptimized version). With the bottleneck out of the way, the speedup curve of Fig. 8 is much improved, making 20-way multiprocessing and beyond a realistic option from the algorithmic point of view.

As explained in Section III, technology constraints limit the size of a tile. In particular, with today's 90 nm technology, a tile with 20 Trimedias and associated memory is not technically feasible. A much more realistic choice is 8 Trimedias per tile. Therefore the unoptimized version of the multi-threaded MPEG2 decoder already provides sufficient parallelism for contemporary CAKE instances.

VIII. CONCLUSIONS

A configurable homogeneous multiprocessor is an extremely versatile chip that can be used to host a wide range of applications for the digital consumer, telecom terminals, networking and ASIC markets. Using YAPI as a modeling language, many different applications can be mapped onto the exact same configurable chip.

There are many advantages to such a platform, including high computational efficiency and very high performance levels, full programmability for a short time-to-market, no mask costs, and no test chip tape-outs until a product succeeds in a high volume market, and high yield because partially defective die are tolerated.

ACKNOWLEDGMENT

Special thanks to Erwin de Kock and Willem Mallon of Philips Research for their discussion on the merits of Kahn process networks versus dynamic data flow networks.

REFERENCES

1. Garcia-Molina, H., Rogers, L. R. (1987). Performance through memory. In: *Proceedings of the ACM Conference on Measurement and Modeling of Computer Systems,* pp. 122–131, May.
2. TM32A data sheet, www.trimedia.com
3. de Kock, E. A., Essink, G., Smits, W. J. M., van der Wolf, P., Brunel, J.-Y., Kruijtzer, W. M., Lieverse, P., Vissers, K. A. (2000). YAPI: Application Modeling for Signal Processing Systems. In: *Proceedings of the 37th Design Automation Conference,* Los Angeles.
4. Kahn G. (1974). The semantics of a simple language for parallel programming. In: Rosenfeld, J. L., ed. *Information Processing*. North-Holland Publishing Company pp. 471–474.
5. Basten, T. Hoogerbrugge, J. (2001). Efficient Execution of Process Networks. In: *Proceedings of Communicating Process Architectures,* Bristol, UK, September.
6. von Eicken, T., Culler, D., Goldstein, S., Schauser, K. (1992). Active Messages: a Mechanism for Integrated Communication and Computation. In: *Proceedings of the 19th International Symposium on Computer Architecture,* pp. 256–266, May.
7. Haskell, B.G., Puri, A, Netravali, A.N. (1997). *Digital video: an introduction to MPEG-2*. Chapman & Hall.

12

Future Directions of Programmable and Reconfigurable Embedded Processors

Stephan Wong, Stamatis Vassiliadis, and Sorin Cotofana
Delft University of Technology, Delft, The Netherlands

I. INTRODUCTION

A technology turning point that made embedded consumer electronics systems an everyday reality had to be the advent of microprocessors. The technological developments that allowed single-chip processors (microprocessors) made embedded systems inexpensive and flexible. Consequently, microprocessor-based embedded systems were introduced into many new application areas. Currently, embedded programmable microprocessors, in one form or another (from 8-bit micro-controllers to 32-bit digital signal processors and 64-bit RISC processors), are everywhere in consumer electronic devices, home appliances, automobiles, network equipment, industrial control systems, and so on. Interestingly, we are utilizing more than several dozens of embedded processors in our day-to-day lives without actually realizing it. For example, in modern cars such as the Mercedes S-class or the BMW 7-series, we can find over 60 embedded processors that control a multitude of functions (e.g., the fuel injection and the anti-lock braking system [ABS], that guarantee a smooth and foremost safe drive). Furthermore, the employment of embedded microprocessors appears to grow in an exponential curve.

In this chapter, we describe several characteristics of embedded processors and investigate how these characteristics have changed over time driven by market requirements such as smaller time-to-market windows and reduced development costs. Subsequently, we discuss two widely employed strategies to meet such market requirements, namely programmability and reconfigurability. Finally, we present a possible future direction in embedded processor design that merges both strategies and thereby providing flexibility in both software and hardware design at the same time.

This chapter is organized as follows. Section II introduces a definition of embedded systems, discusses the ensuing characteristics of embedded systems, and provides an in-depth discussion of traditional embedded processor characteristics. Section III discusses the need for programmability and several examples of such an approach. Section IV discusses how the use of reconfigurability affected the embedded processor characteristics. Section V describes a possible future direction in embedded processor design that combines programmability and reconfigurability. Furthermore, we show an example of such an approach called the microcoded reconfigurable embedded processor, also called the MOLEN processor. Section VI concludes by stating several key observations discussed in this chapter.

II. TRADITIONAL EMBEDDED PROCESSOR CHARACTERISTICS

An embedded processor is a specific instance of embedded systems in general and therefore adhere to the characteristics of embedded systems. Since no generally accepted definition of embedded systems exists, we establish our own definition in order to facilitate the discussion on embedded system characteristics and subsequently on embedded processor design issues.

> *Definition*: Embedded systems are (inexpensive) mass-produced elements of a larger system providing a dedicated, possibly time-constrained, service to that system.

Before we highlight the main characteristics of embedded systems, we would like to comment on our one-sentence definition of them. In most literature, the definition of embedded systems only states that they provide a dedicated service—the nature of the service is not relevant in this context—to a larger (embedding) system. Consequently, when we refer to embedded systems as mass-produced elements, we draw the separation line between application-specific systems and embedded systems. We are aware that the separation line is quite thin in the sense that embedded systems are mostly indeed application-specific systems. However, we believe that application-specific systems produced in low volumes can not be considered to be embedded systems, because they represent a niche market for which completely different requirements are valid. For example, cost is unimportant in a low-volume production scenario contrary to the paramount importance to achieve low cost for embedded systems. Finally, we include the possibility for time-constrained behavior in our definition, because even though it is not characteristic to all embedded systems it constitutes a particularity of a very large class of them, namely real-time embedded systems.

Clearly, the precise requirements of an embedded system are determined by its immediate environment. The immediate environment of an embedded system can be either other surrounding embedded systems in the larger embedding system or even the world in which the larger system is placed. We can classify the embedded system requirements into:

Functional requirements are defined by the services that the embedded system must perform for and when interacting with its immediate environment. Such services possibly include data gathering and exerting control to their immediate environment. This implies that some kind of data transformation must be performed within the embedded system itself.

Temporal requirements are the result of the time-constrained behavior of many embedded systems, thereby introducing deadlines (explained later) for the service(s).

Dependability requirements relate to the reliability, maintainability, and availability of the embedded system in question.

In the light of the previously stated embedded systems definition and requirements, we briefly point out what we think are the main characteristics of more traditional embedded processors. Furthermore, we discuss in more detail the implications that these characteristics have on the specification and design processes of embedded processors. The first and probably the most important characteristic of embedded processors is that they are *application-specific*. Given that the service (or application in processor terms) is known a priori, the embedded processor can be and should be optimized for its targeted application. In other words, embedded processors are definitely not general-purpose processors that are designed to perform reasonably for a much wider range of applications. Moreover, the fact that the application is known beforehand opens the road for *hardware/software co-design* (i.e., the cooperative and concurrent design of both hardware and software components of the processor). The hardware/software co-design style is very much particular to embedded processors and has the goal of meeting the processor level objectives by exploiting the synergism of hardware and software.

Another important characteristic of embedded processors is their *static structure*. When considering an embedded processor, the end-user has very limited access to programming. Most of the software is provided by the processor integrator and/or application developer, reside on ROM memories, and run without being visible to the end-user. The end-user can neither change nor reprogram the basic operations of the embedded processor, but is usually allowed to program the embedded system by re-arranging the sequence of basic operations.

Embedded processors are essentially nonhomogeneous processors and this characteristic is induced by the *heterogeneous* character of the process within which the processor is embedded. Designing a typical embedded processor does not only mix hardware design with software design, but it also mixes design styles within each of these categories. To put more light on the heterogeneity issue, we depicted in Fig. 1 an example of a signal processing embedded processor. The heterogeneous character can be seen in many aspects of the embedded processor design as follows.

Both analog and digital circuits may be present in the system.
The hardware may include microprocessors, microcontrollers, digital signal processors, and application-specific integrated circuits.
The topology of the system is rather irregular.
The software may include various software modules as well as a multitasking real-time operating system.

Generally speaking, the intrinsic heterogeneity of embedded processors largely contributes to the overall complexity and management difficulties of the design process. However, one can say that heterogeneity is, in the case of embedded processor design, a necessary evil. It provides better design flexibility by providing a wide range of design options. In addition, it allows

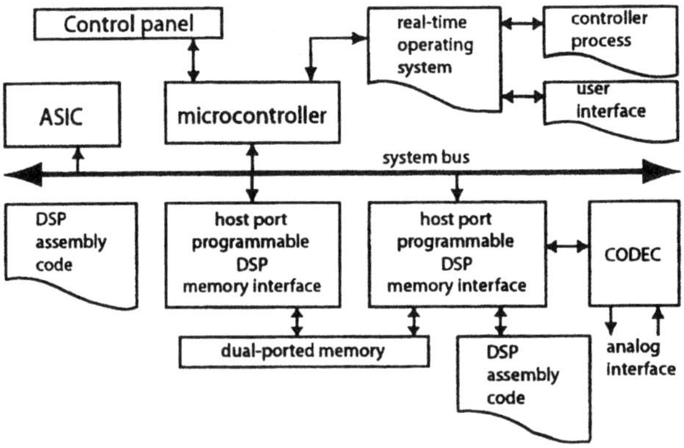

Figure 1 Signal processing embedded processor example. (From Ref. 7.)

each required function to be implemented on the most adequate platform that is deemed necessary to meet the posed requirements.

Embedded processors are *mass-produced* application-specific elements separating them from other low-volume produced application-specific processors. Embedded processors represent a much larger market segment in which embedded processor vendors face fierce competition in order to gain market capitalization. Consequently, this environment imposes a different set of requirements on the embedded processor design. For example, such requirements involve the cost/performance sensitivity of embedded processors and make low cost almost always an issue.

A large number of embedded processors perform *real-time* processing, introducing the notion of *deadlines*. Roughly speaking, deadlines can be classified into hard and soft real-time deadlines. Missing a hard deadline can be catastrophic, while missing a soft deadline only results in nonfatal glitches at most. Both types of deadlines are known a priori much like the functionality is known beforehand. Therefore, deadlines determine the minimum level of performance that must be achieved. When facing hard deadlines, special attention must also be paid to other components within the larger embedding system that are connected to the embedded processor in question since they can negatively influence its behavior.

III. THE NEED FOR PROGRAMMABILITY

In the early 1990s, we witnessed a trend in the embedded processor market that was reshaping the characteristics of traditional embedded processors as introduced in Section II. Driven by market forces, the lengthy embedded processor design cycles had to be shortened in order to keep up with or stay in front of competitors and costs had to be reduced in order to stay competitive. More specifically, the cost of an embedded processor can be largely divided into production costs (closely related to the utilized manufacturing technology) and development costs (closely related to overall design cycle). It must be clear that the production costs remain constant for each produced embedded processor due to the fact that the embedded processor design must be fixed before entering production. Since we focus on embedded processor design and not on manufacturing, the issues concerning production costs are left out of the ensuing discussion. However, we must note that the complexity of the final embedded processor design certainly has an impact on production costs. The impact is exhibited by requiring more steps in the manufacturing process and/or a more expensive manufacturing process altogether. On the other hand, the development costs on a per embedded processor basis can be reduced by amortizing the costs over a higher production volume. Certainly, this greatly depends on the market demand and the established market capitalization. Alternatively and maybe more beneficial is to reduce the design cycle and therefore its associated costs altogether. In this section, by highlighting the traditional embedded processors design, we discuss "large scale" programmability that has been used to address the issues of lengthy design cycles and the associated development costs. One could argue that programmability has always been part of embedded processors. However, programmability introduced in this section significantly differs from the limited (low-level) programmability of traditional embedded processors.

The heterogeneity of embedded systems demands a multitude of embedded processors to be designed for a single system. This was further strengthened by the fact that semiconductor technology at the time did not allow large chips to be manufactured. Subsequently, the design of embedded processors required lengthy design cycles and especially lengthy verification cycles for the chips and their interfaces. On the other hand, one can argue that an advantage is that subsequent system design cycles could significantly be reduced to only one or a few embedded processors that needed to be redesigned. This delicate balance between long *initial design cycles* and possibly shortened *subsequent design cycles* was disturbed when advances in semiconductor technology allowed increasingly more gates to be put on a single chip. As a result, more functionality migrated from a multitude of embedded processors into a single one. The resulting design of more complex and larger embedded processors did not have a great effect on the initial design cycles. However, the length of subsequent redesign cycles increased since the utilization of optimized circuits meant that subsequent designs were not necessarily easier than the initial ones.

In the search for design flexibility in order to decrease design cycles and reduce subsequent design costs, functions were separated into time-critical functions and non-time-critical ones. The embedded processors design paradigm had shifted from one that was based on the functional requirements to one that is based on the temporal requirements. The collection of non-time-critical functions could then be performed by a single chip (possibly implemented in a slower technology in order to reduce cost). The remaining time-critical functions are to be implemented in high-speed circuits achieving maximum performance. The main benefit of this approach is that the larger and (possibly) slower chips can be reused in subsequent designs resulting in shorter subsequent design cycles. While this design paradigm was born out of market needs (i.e., to reduce design cycles and development costs), it is well-known in the design of general-purpose purpose processors. In the general-purpose processor design paradigm, the processor design can be divided into three distinct fields architecture,* implementation, and realization [5].

In Section II, we stated that more traditional embedded processors were application-specific and static in nature. However, in this section we also stated that increasingly more functionality is embedded into a single embedded processor. Is such a processor still application-specific and can we still call such a processor an embedded processor? The answer to this question is affirmative since such a processor is still embedded if the other constraints (mass-produced, providing a dedicated service, etc.) are observed. Given that increasing functionality usually implies more exposure of the processor to the programmer, embedded processors have indeed become less static as they can now be reused for other application areas due to their programmability. In light of this all, two scenarios in the design of programmable embedded processors can be distinguished:

1. *Adapt an existing general-purpose architecture and implement it.* This scenario reduces development costs albeit such architectures must usually be licensed. Furthermore, since such architectures were not adapted to embedded processors, some development time is still needed to modify such architectures.
2. *Build a new embedded processor architecture from scratch.* In this scenario, the embedded processor development takes longer, but the final architecture is more focused on the targeted application(s) and thus possibly achieves better performance than already existing general-purpose architectures. Actually, the goal is to develop an architecture for a collection of similar applications (called application domain) such that processors can be produced once and reused when placed in different environments. This reduces the overall system cost since the development costs are amortized over a higher number of embedded processors.

Several examples of the first scenario can be found. A well-known example is the MIPS architecture [13], which has been adapted resulting in several embedded processor families. In this case, the architecture has been increasingly adapted toward embedded processors by MIPS Technologies, Inc., which develops the architecture independently from other embedded systems vendors. Another well-known example is the ARM architecture [21] found in many current embedded processors. It is a RISC architecture that was intended for low-power PCs (1987) at first, but it has been quickly

* The architecture of any computer system is defined as the conceptual structure and functional behavior as seen by its immediate user.

adapted to become an embeddable RISC core (1991). Since then the architecture was modified and extended several times in order to optimize it for its intended applications. The most well-known version is the StrongARM core, which was jointly developed by Digital Semiconductor and ARM. This core was intended to provide great performance at an extreme low-power. The most recent and extended implementation of this architecture was developed by Intel called the Intel PCA Application Processor [14]. Other examples of general-purpose architecture that have been adapted include: IBM PowerPC [16], Sun UltraSPARC [25], and the Motorola 68k/Coldfire [18]. An example of the second scenario is the Trimedia VLIW architecture [22] from Trimedia Technologies, Inc., which was originally developed by Philips Electronics, N.V. Its application domain is multimedia processing and processors based on this architecture can be found in television sets, digital receivers, and other digital video editing boards. It contains a VLIW processor core that performs non-time-critical functions and also controls the specialized hardware units that are intended for specific real-time multimedia processing.

Summarizing, the characteristics mentioned in Section II can be easily reflected in the three design stages: architecture, implementation, and realization. The characteristic of embedded processors being application-specific processors is exhibited by the fact that the architecture only contains those instructions that are really needed to support the application domain. The static structure characteristic exhibits itself by having a fixed architecture, a fixed implementation, and a fixed realization. The heterogeneity characteristic exhibits itself by the utilization of a programmable processor core with other specialized hardware units. Such specialized hardware units can possibly be implemented on the same chip as the programmable processor core. Extending this principle further, the heterogeneity of the embedded processor also exhibits itself in the utilization of different functional units in the programmable processor core. The mass-produced characteristic is exhibiting itself in the realization process by only utilizing proven technology that therefore should be available, cheap, and reliable. The requirement of real-time processing exhibits itself by requiring architectural support for frequently used operations, extensively parallel and/or pipelined (if possible) implementations, and realizations incorporating adequately high-speed components.

IV. EARLY TIME RECONFIGURABILITY

In the mid-1990s, we witnessed a second trend in the embedded processors design next to programmability that was likewise reshaping the design methodology of embedded processors and consequently redefined some of their characteristics. Traditionally, the utilization of application-specific integrated circuits (ASICs) was commonplace in the design of embedded processors resulting in lengthy design cycles. Such an approach requires several roll-outs of embedded processor chips in order to test/verify all the functional, temporal, and dependability requirements. Therefore, design cycles of 18 months or longer were commonplace rather than exceptions. A careful step toward reducing such lengthy design cycles is to utilize reconfigurable hardware, also referred to as fast prototyping. The utilization of reconfigurable hardware allows embedded processor designs to be mapped early on in the design cycle to reconfigurable hardware, in particular field-programmable gate arrays (FPGAs), giving rise to three advantages. First, the mapping requires considerably less time than a chip roll-out and thereby shortens the development time. Second, the embedded processor functionality can be tested in an earlier stage and allows more design alternatives to be explored. Third, the number of (expensive) chip roll-outs is also reduced and thereby further reduces the development costs. However, the reconfigurable hardware was initially limited in size and speed. The limited size meant that only partial designs could be tested. Consequently, roll-out of the complete embedded processor design (implemented in ASICs) were still required in order to verify the overall functionality and performance.

In recent years, reconfigurable hardware technology has progressed at a fast pace, arriving at the point where embedded processor designs requiring millions of gates can be implemented on such structures. Moreover, the existing performance gap betwen FPGAs and ASICs is rapidly decreasing. Due to these technological developments, the role of reconfigurable hardware in embedded processor design has changed considerably. In the following paragraphs, we revisit the traditional embedded processor characteristics mentioned in Section II and investigate whether they still hold for the case of FPGA-based embedded processors.

Application-specific Embedded processors built utilizing reconfigurable hardware are still application-specific in the sense that the implementations are still targeting such applications. Utilizing such implementations for other purposes will prove to be very difficult or even impossible, because the required performance levels most certainly can not be achieved.

Static structure From a pure technical perspective, the structure of a reconfigurable embedded processor is not static since its functionality can be changed during its lifetime. However, in most cases the design implemented in reconfigurable hardware remains fixed between maintenance intervals. Therefore, from the user's perspective the structure of the embedded processor is still static. In the next section, we explore the possibility that the functionality of an embedded processor needs to be changed even during run-time. In this case, the static structure can be perceived from a higher perspective, namely the reconfigurable hardware is designed to support only a fixed (or static) set of implementations.

Heterogeneous This characteristic is still very much present in the case of reconfigurable embedded processors. We have added additional technology into the mix in which embedded processors can be realized. For example, the latest FPGA offering from both Altera, Inc. (Stratix [2]) and Xilinx, Inc. (Virtex II [29]) integrates memory, logic, I/O controllers, and DSP blocks on a single chip.

Mass-produced This characteristic is still applicable to current reconfigurable embedded processors. Early on, reconfigurable hardware was expensive resulting in its sole utilization for fast prototyping purposes. As the technology progressed, reconfigurable hardware became cheaper and this opened the possibility of actually shipping reconfigurable embedded processors in final products. An important enabling trend (next to reduced cost) that must not be overlooked is that reconfigurable hardware has also become more reliable both in production and during operation.

Real-time In the beginning, we witnessed the incorporation of reconfigurable hardware only for non-'time-critical' functions. As the technology of reconfigurable hardware continues to progress and make reconfigurable hardware much faster, we are also witnessing their incorporation in actual products where real-time performance is required, such as multimedia decoders.

V. FUTURE EMBEDDED PROCESSORS

In Sections III and IV, we argued that both programmability and reconfigurability have been introduced into the embedded processor design trajectory born out of the need to reduce design cycles and reduce development costs. In short, programmability allows the utilization of high-level programming languages (like C) and makes it easier to support applications on embedded processors. Reconfigurability allows designs to be tested early on in terms of functionality and diminishes the need for expensive chip roll-outs. Merging both strategies is a logical and evolutionary step in embedded processor design and has enormous potential, especially when considering that the performance of FPGA is nearing that of ASIC. More precisely, we believe that the merging encompasses the augmentation of a programmable processor (core) with reconfigurable hardware, possibly replacing fixed

(ASICs) hardware. We foresee that such an augmentation will provide several advantages:

Improved performance compared to a software-only implementation, because (tuned) specialized hardware implemented on the FPGA can exploit the parallelism of the supported function and allow the utilization of other performance-increasing techniques.

Rapid application development since the mentioned augmentation introduces the possible utilization of high-level programming and high-level hardware description languages in the design trajectory.

Design flexibility is achieved by allowing design space exploration in both hardware and software due to the possible utilization of high-level hardware description programming and hardware description languages.

These advantages and enabling FPGA technologies have even resulted in programmable processor cores that are under consideration to be implemented in the same FPGA structures (e.g., Nios from Altera [1] and MicroBlaze from Xilinx [30]). However, the utilization of programmable embedded processors that are augmented with reconfigurable hardware also poses several issues that must be addressed:

Long reconfiguration latencies. In run-time reconfiguration, such latencies may greatly penalize the performance, because any computation must be halted until the reconfiguration has finished.

Limited opcode space. The initiation and control of the reconfiguration and execution of various implementations on the reconfigurable hardware require the introduction of new instructions. This puts much strain on the opcode space.

Complicated decoder hardware. The multitude of new instructions greatly increases the complexity of the decoder hardware.

In the following, we introduce and discuss one possible approach [24] in merging programmability with reconfigurability in the design of embedded processors. The approach utilizes microcode to alleviate the above-mentioned problems. Microcode consists of a sequence of (simple) micro-instructions that, when executed in a certain order, performs "complex" operations. This approach allows "complex" operations to be performed on much simpler hardware. In this section, we consider the reconfiguration (either off-line or run-time) and execution processes as complex operations. The main benefits of our approach can be summarized as follows:

Reduced reconfiguration latencies. Microcode used to control the reconfiguration process allows itself to be cached on-chip. This results in faster access times to the reconfiguration microcode and thus in turn reduces the reconfiguration latencies.

Reduced opcode space requirements. By only pointing to microcode (explained later), we only require (at most) three new instructions to support any current and future operations.

Reduced decoder complexity. By introducing only a few instructions, no complex instruction decoding hardware is required.

In Section A, we revisit microcode from its beginnings to its current implementation within a high-level microprogrammed machine. In Section B, we discuss in-depth our proposed MOLEN embedded processor. Finally, in Section C we briefly highlight several other approaches in this field that are comparable in one way or another.

A. Revisiting Microcode

Microcode, introduced in 1951 by Wilkes [26], constitutes one of the key computer engineering innovations. Microcode de facto partitioned computer engineering into two distinct conceptual layers, namely, architecture and implemention. This is in part because emulation allowed the definition of complex instructions that may have been technologically not implementable (at the time they were defined), thus projecting an architecture to the future. That is, it allowed computer architects to determine a technology-independent functional behavior (e.g., instruction set) and conceptual structures providing the following possibilities:

Define the computer's architecture as a programmer's interface to the hardware rather than to a specific technology dependent realization of a specific behavior.

Allow a single architecture to be determined for a "family" of implementations giving rise to the concept of compatibility. Simply stated, it allowed programs to be written for a specific architecture once and run at "infinitum" independent of the implementations.

Since its beginning, as introduced by Wilkes, microcode has been a sequence of micro-operations (microprograms). Such a microprogram consists of pulses for operating the gates associated with the arithmetical and control registers. Figure 2 depicts the method of generating this sequence of pulses. First, a timing pulse initiating a micro-operation enters the decoding tree and, depending on the setup register R, an output is generated. This output signal passes to matrix A, which in turn generates pulses to control arithmetical and control registers, thus performing the required micro-operation. The output signal also passes to matrix B, which in its turn

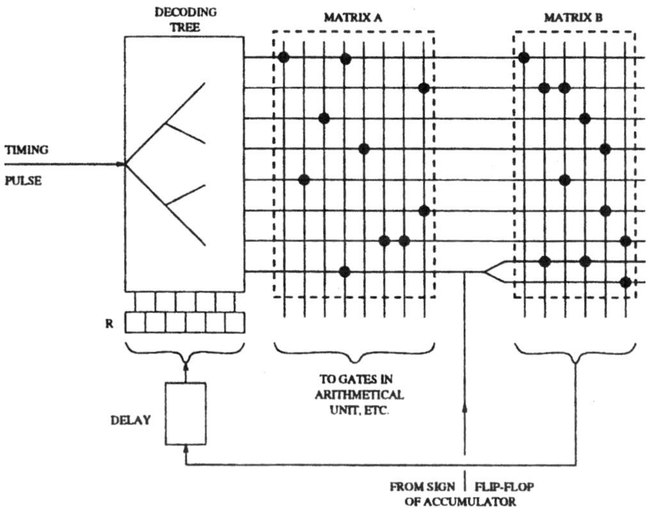

Figure 2 Wilkes' microprogram control model [26].

generates pulses to control the setup register R (with a certain delay). The next timing pulse will therefore generate the next micro-operation in the required sequence due to the changed register R.

Over the years, Wilkes's model has evolved into a high-level microprogrammed machine as depicted in Fig. 3. In this figure, the memory address register (MAR) is used to store the memory address in the main memory from which data must be loaded and to which data is stored. The memory data register (MDR) stores the data that is communicated to or from the main memory. Furthermore, the control store contains microinstructions (representing one or more micro-operations) and the sequencer determines the next microinstruction to execute. The control store and the sequencer correspond to Wilkes's matrices A and B respectively. The machine's operation is as follows:

1. The control store address register (CSAR) contains the address of the next microinstruction located in the control store. The microinstruction located at this address is then forwarded to the microinstruction register (MIR).
2. The MIR decodes the microinstruction and generates smaller micro-operation(s) accordingly that need to be performed by the hardware unit(s) and/or control logic.
3. The sequencer utilizes status information from the control logic and/or results from the hardware unit(s) to determine the next

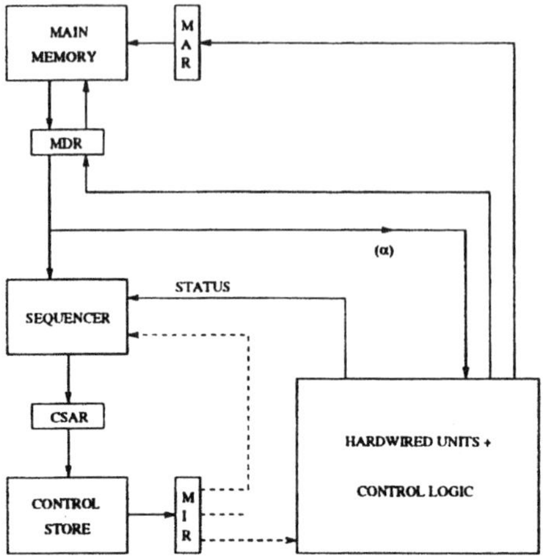

Figure 3 A high-level microprogrammed machine.

microinstruction and stores its control store address in the CSAR. It is also possible that the previous microinstruction influences the sequencer's decision regarding which microinstruction to select next.

It should be noted that in microcoded engines not all instructions access the control store. As a matter of fact, only emulated instructions have to go through the microcode logic. All other instructions will be executed directly by the hardware (following path (α) in Fig. 3). That is, a microcoded engine is as a matter of fact a hybrid of the implementation having emulated instructions and hardwired instructions. We must note that contrary to some beliefs, from the moment it was possible to implement instructions, microcoded engines have always had a hardwired core that executes RISC instructions.

B. Microcoded Reconfigurable MOLEN Embedded Processor

In this section, only a brief description of the MOLEN embedded processor is given. For a more detailed description we refer to [24,28]. In its most general form, the proposed machine organization augmented with a reconfigurable unit is depicted in Fig. 4. In this organization, instructions are fetched from the main memory and are temporarily stored in the "instruction fetch" unit. Subsequently, these instructions are fetched by the "arbiter," which decodes them before issuing them to their corresponding

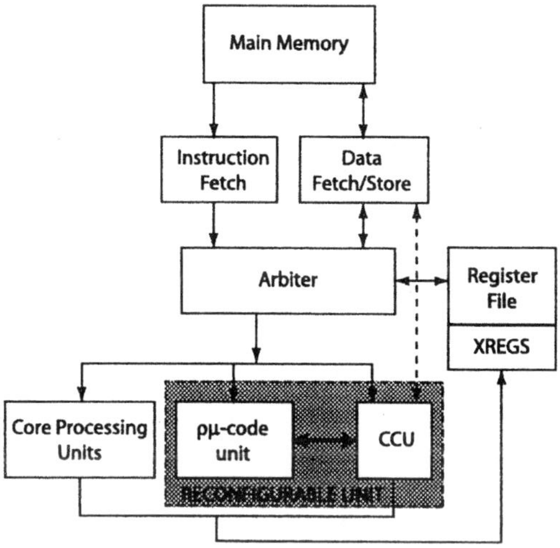

Figure 4 The MOLEN machine organization.

execution units. Instructions that have been implemented in fixed hardware are issued to the "core processing units" (i.e., the regular functional units such as ALUs, multipliers, and divider). Instructions that have been implemented in reconfigurable hardware are issued to the "reconfigurable unit." Similar to other load/store architectures, the proposed machine organization executes data that is stored in the register file and prohibits direct memory data accesses by hardware units other than the load/store unit(s). However, there is one exception to this rule, the custom configured unit (CCU), which embodies the actual reconfigurable hardware (e.g., FPGA), is also allowed direct memory data access via the "data fetch/store" unit (represented by a dashed two-ended arrow). This enables the CCU to perform much better when streaming data accesses are required (e.g., in multimedia processing). Finally, we introduced the exchange registers (XREGS) which are utilized to accommodate a more extensive argument passing mechanism (compared to registers which are restricted in number and size) between the complex implementations configured on the CCU and the application code which embeds such implementations.

The reconfigurable unit consists of a custom configured unit (CCU), which could be for example be implemented by a field-programmable gate array (FPGA), and the ρμ-code unit. An operation, which can be as simple as an instruction or as complex as a piece of code, performed by the reconfigurable unit is divided into two distinct process phases: `set` and `execute`. The `set` phase is responsible for configuring the CCU enabling it to perform the required operation(s). Such a phase may be subdivided into two sub-phases: partial set (*p*-set) and complete set (*c*-set). The *p*-set sub-phase is envisioned to cover common functions of an application or set of applications. More spefically, in the *p*-set sub-phase the CCU is *partially* configured to perform these common functions. While the p-set sub-phase can possibly be performed during the loading of a program or even at chip fabrication time, the *c*-set sub-phase is performed during program execution. In the *c*-set sub-phase, the remaining part of the CCU (not covered in the *p*-set sub-phase) is configured to perform other less common functions and thus *completing* the functionality of the CCU. The configuration of the CCU is performed by executing reconfiguration microcode (either loaded from memory or resident) in the ρμ-code unit. Reconfiguration microcode is generated by translating a reconfiguration file into microcode. In the case that partial reconfigurability is not possible or not convenient, the *c*-set sub-phase can perform the entire configuration. The *execute* phase is responsible for actually performing the operation(s) on the (now) configured CCU by executing (possibly resident) execution microcode stored in the $\rho\mu$-code unit.

In relation to these three phases, we introduce three new instructions: *c*-set, *p*-set, and execute. Their instruction format is given in Fig. 5. We must note that these instructions do *not* specifically specify an operation and then load the corresponding reconfiguration and execution microcode. Instead, the *p*-set, *c*-set, and execute instructions directly point to the (memory) location where the reconfiguration or execution microcode is stored. In this way, different operations are performed by loading different reconfiguration and execution microcodes. That is, instead of specifying new instructions for the operations (requiring instruction opcode space), we simply point to (memory) addresses. The location of the microcode is indicated by the resident/pageable-bit (R/P-bit) which implicitly determines the interpretation of the address field (i.e., as a memory address α (R/P = 1) or as a ρ-CONTROL STORE address ρCS-α (R/P = 0) indicating a location within the $\rho\mu$-code unit). This location contains the first instruction of the microcode which must always be terminated by an *end_op* microinstruction.

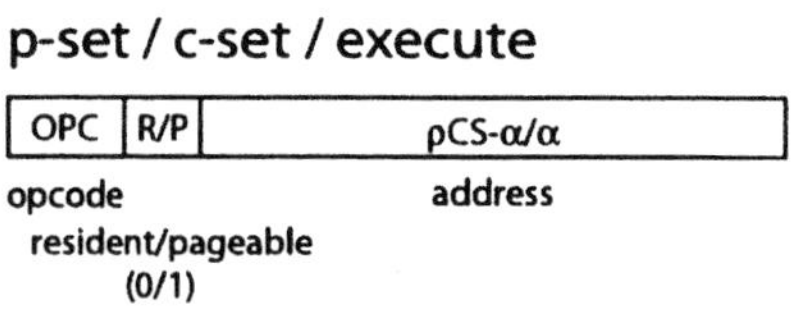

Figure 5 The *p*-set *c*-set, and execute instruction formats.

The ρμ-code unit. The $\rho\mu$-code unit can be implemented in configurable hardware. Since this is only a performance issue and not a conceptual one, it is not considered in further detail. In this presentation, for simplicity, we assume that the ρμ-code unit is hardwired. The internal organization of the $\rho\mu$-code unit is given in Fig. 6. In all phases, microcode is used to perform either reconfiguration of the CCU or control the execution on the CCU. Both types of microcode are conceptually the same and no distinction is made between them in the remainder of this section. The $\rho\mu$-code unit comprises two main parts: the SEQUENCER and the ρ-CONTROL STORE. The SEQUENCER mainly determines the microinstruction execution sequence and the ρ-CONTROL STORE is mainly used as a storage facility for microcodes. The execution of microcodes starts with the SEQUENCER receiving an address from the ARBITER and interpreting it according to the R/P-bit. When receiving a memory address, it must be determined whether the microcode is already cached in the ρ-CONTROL STORE or not. This is done by checking the RESIDENCE TABLE which stores the most frequently used translations of memory addresses into ρ-CONTROL STORE addresses and keeps track of the validity of

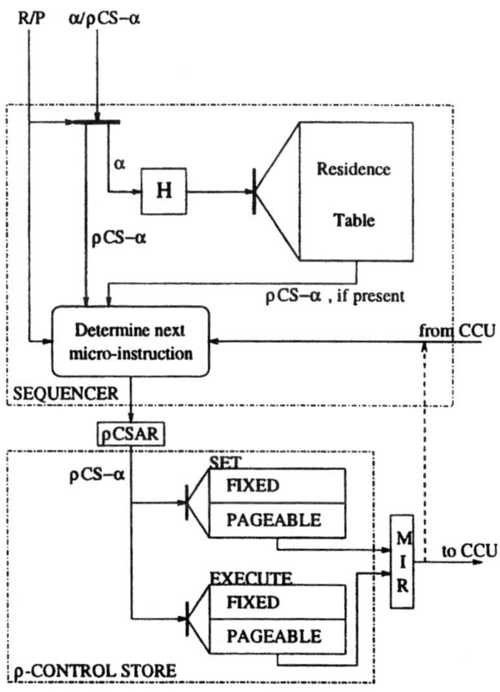

Figure 6 $\rho\mu$-code unit internal organization.

these translations. It can also store other information: least recently used (LRU) and possibly additional information required for virtual addressing support. In the remainder we assume that the system only allows for real addressing for simplicity of discussion. In the cases that a ρCS-α is received or a valid translation into a ρCS-α is found, it is transferred to the "determine next microinstruction" block. This block determines which (next) microinstruction needs to be executed.

When receiving address of first microinstruction: Depending on the R/P-bit, the correct ρCS-α is selected, i.e., from instruction field or from RESIDENCE TABLE.

When already executing microcode: Depending on previous microinstruction(s) and/or results from the CCU, the next microinstruction address is determined.

The resulting ρCS-α is stored in the ρ-control store address register (ρCSAR) before entering the ρ-CONTROL STORE. Using the ρCS-α, a microinstruction is retrieved from the ρ-CONTROL STORE and then stored in the microinstruction register (MIR) before it controls the CCU reconfiguration or before it is executed by the CCU.

The ρ-CONTROL STORE comprises two section,* namely a `set` section and an `execute` section. Both sections are further divided into a `fixed` part and `pageable` part. The fixed part stores the resident reconfiguration and execution microcode of the `set` and `execute` phases, respectively. Resident microcode is commonly used by several invocations (including reconfigurations) and it is stored in the fixed part so that the performance of the `set` and `execute` phases is possibly enhanced. Which microcode resides in the fixed part of the ρ-CONTROL STORE is determined by performance analysis of various applications and by taking into consideration various software and hardware parameters. Other microcodes are stored in memory and the pageable part of the ρ-CONTROL STORE acts like a cache to provide temporal storage. Cache mechanisms are incorporated into the design to ensure the proper substitution and access of the microcode present in the ρ-CONTROL STORE.

C. Other Reconfigurability Approaches

In the previous subsection, we introduced a machine organization where the hardware reconfiguration and the execution of the reconfigured hardware was done in firmware via the ρ-microcode (an extension of the classical microcode to include reconfiguration and execution for resident and non-resident

* Both sections can be identical, but are probably only differing in microinstruction word sizes.

microcode). The microcode engine was extended with mechanisms that allow for permanent and pageable reconfiguration and execution microcode to coexist. We also provided partial reconfiguration possibilities for "off-line" configuration and prefetching of configurations. Regarding related work, we considerd more than 40 machine proposals. We report here a number of them that somehow use some partial or total reconfiguration prefetching. It should be noted that our scheme is rather different in principle from all related work as we use microcode, pageable/fixed local memory, hardware assists for pageable reconfiguration, partial reconfigurations, etc. As it will be clear from the short description of the related work, we differentiated from them in one or more mechanisms.

The programmable reduced instruction set computer (PRISC) [19] attaches a programmable functional unit (PFU) to the register file of a processor for application-specific instructions. Reconfiguration is performed via exceptions. In an attempt to reduce the overhead connected with FPGA reconfiguration, Hauck proposed a slight modification to the PRISC architecture in [11]: an instruction is explicitly provided to the user that behaves like a NOP if the required circuit is already configured on the array, or is in the process of being configured. By inserting the configuration instruction before it is actually required, a so-called *configuration prefetching* procedure is initiated. At this point the host processor is free to perform other computations, overlapping the reconfiguration of the PFU with other useful work. The *OneChip* introduced by Wittig and Chow [27] extends PRISC and allows the PFU for implementing any combinational or sequential circuits, subject to its size and speed. The system proposed by Trimberger [23] consists of a host processor augmented with a PFU, reprogrammable instruction set accelerator (RISA), much like the PRISC mentioned above. Concerning the management and control of the reprogramming procedure, Trimberger mentions that the RISA reconfiguration is under control of a hardwired execution unit. However, it is not obvious if an explicit SET instruction is available. The reconfigurable multimedia array coprocessor (REMARC) proposed by Miyamori and Olukotun [17] augments the instruction set of a MIPS core. As the coprocessor does not have a direct access to the main memory, the host processor must write the input data to the coprocessor data registers, initiate the execution, and finally read the results from the coprocessor data registers. An explicit reconfiguration instruction is provided. *Garp*, designed by Hauser and Wawrzynek [12], is another example of a MIPS-derived custom computing machine (CCM). The FPGA-based coprocessor has direct access to the standard memory. The MIPS instruction set is augmented with several nonstandard instructions dedicated to loading a new configuration, initiating the execution of the newly configured computing facilities, moving data between the array and the processor's own registers, saving/retrieving the array states, branching on conditions provided by the array and so on. The coprocessor is aimed to run autonomously with the host processor. In the *OneChip*-98 introduced by Jacob and Chow [15], the computing resources are loaded *on demand* when a miss is detected. Alternatively, the resources are preloaded by using compiler directives. Several comments regarding these assertions are noteworthy. If an on-demand loading strategy is employed, then the user has no control on the reconfiguration procedure. In the preloading strategy, an explicit reconfiguration instruction is provided to the user and the reconfiguration procedure is indeed under the control of the user. PRISM (processor reconfiguration through instruction-set metamorphosis) one of the earliest proposed CCM [3,4], was developed as a proof-of-concept system, in order to handle the loading of FPGA configuration, the compiler inserts library function calls into the program stream [4]. From this description, we can conclude that an explicit reconfiguration procedure is available. Gilson [8] CCM architecture consists of a host processor and two or more FPGA-based *computing devices*. The host controls the reconfiguration of FPGA by loading new configuration data through a host interface into the FPGA configuration memory. The reconfiguration process can be performed such that when one computing device is being reconfigured and, therefore, is idle, the others continue executing. The write into the configuration memory instruction can play the role of an explicit reconfiguration instruction. Therefore, a *preloading* strategy is employed. Schmit [20] proposes a partial run-time reconfiguration mechanism, called *pipeline reconfiguration* or *striping*, by which the FPGA is reconfigured at a granularity that corresponds to a pipeline stage of the application being implemented. An application that has been broken up into pipeline stages can be mapped to a striped FPGA. The pipeline stages are known as *stripes*; the stages of the application are called *virtual stripes*; and the hardware stages at which the virtual stages are loaded into are called *physical stripes*. The PipeRench coprocessor developed by a team with Carnegie Mellon University [6,9] focused on implementing linear (1-D) pipelines of arbitrary length. PipeRench is envisioned as a coprocessor in a general-purpose computer, and has direct access to the same memory space as the host processor. The virtual stripes of the application are stored into an on-chip configuration memory. A single physical stripe can be configured in one read cycle with data stored in such a memory. The configuration of a stripe takes place concurrently with execution of the other stripes. The reconfigurable data path architecture (rDPA) is also a self-steering autonomous reconfigurable architecture. It consists of a mesh of identical data path units (DPU) [10]. The data-flow direction through the mesh is only from west and/or north to east and/or south and is also data-driven. A word entering rDPA contains a configuration bit which is used to distinguish the configuration information from data. Therefore, a word can specify either a SET or an EXECUTE instruction, the arguments of the instructions being the configuration information or data to be processed. A set of computing facilities can be configured on rDPA.

VI. CONCLUSIONS

In this chapter we described several characteristics of embedded processors that were logically deduced from characteristics of embedded systems in general. Driven by market requirements, two strategies were followed in order to reduce design cycles and development costs. First, programmability was introduced as a means to combine all non-time-critical functions to be performed by a "general purpose" -like embedded processor. Such an embedded processor could then be reused in subsequent designs and, thereby, greatly reduce design cycles. Second, reconfigurability was initially only utilized for fast prototyping. Over time, technological advances in reconfigurable hardware in terms of size and performance have led to the fact the reconfigurable embedded processors are actually incorporated in shipped embedded systems. We believe that the future of embedded processor design lies in the merging of both strategies. Programmability allows the utilization of high-level programming languages (like C) and, thereby, easies application development. The utilization of reconfigurable hardware combines design flexibility and fast prototyping. At the same time, the processing performance of reconfigurable hardware is nearing that of application-specific integrated circuits. Finally, we highlighted one possible framework in which future embedded processor design can be performed. The proposed MOLEN embedded processor combines software programming (by utilizing a programmable processor core) with hardware programming (utilizing microcode to control the reconfigurable hardware). Such an approach provides possibilities in combatting several issues associated with reconfigurable hardware.

REFERENCES

1. Altera Corporation. Nios Embedded Processor. http://www.altera com/products/devices/excalibur/exc-nios_index.html
2. Altera Corporation. Stratix Family. http://www.altera.com/products/devices/stratix/stx-index.jsp
3. Athanas, P.M. (1992). *An Adaptive Machine Architecture and Compiler for Dynamic Processor Reconfiguration*. PhD thesis, Brown University, Providence, Rhode Island, May.

4. Athanas P. M., Silverman H. F. (March 1993). Processor Reconfiguration through Instruction-Set Metamorphosis. *IEEE Computer* 26(3):11–18.
5. Blaauw G. A., Brooks F. P. (1997). *Computer Architecture: Concepts and Evolution*. Addison-Wesley.
6. Cadambi, S., Weener, J., Goldstein, S. C., Schmit, H., Thomas, D. E. (1998). Managing Pipeline-Reconfigurable FPGAs. In: *6th International Symposium on Field Programmable Gate Arrays*. California: USA, pp. 55–64.
7. Chang, W.-T., Kalavade, A., Lee, E. A. (1995). Effective Heterogeneous Design and Co-Simulation. In: de Michelli, G., Sami, M., eds. *Hardware/Software Co-Design*. Kluwer Academic Publishers, pp. 187–211.
8. Gilson, K. L., (1994). Integrated Circuit Computing Device Comprising a Dynamically Configurable Gate Array Having a Microprocessor and Reconfigurable Instruction Execution Means and Method Therefore. U.S. Patent No. 5,361,373, November.
9. Goldstein, S. C., Schmit, H., Moe, M., Budiu, M., Cadambi, S., Taylor, R., Laufer, R. (1999). PipeRench: A Coprocessor for Streaming Multimedia Acceleration. In: *The 26th International Symposium on Computer Architecture*, Georgia, USA, May, pp. 28–39.
10. Hartenstein, R.W., Kress, R., Reinig, H. (1994). A New FPGA Architecture for Word-Oriented Datapaths. In: *4th International Workshop on Field-Programmable Logic: Architectures, Synthesis and Applications*. Lecture Notes in Computer Science. Czech Republic, September, pp. 144–155.
11. Hauck, S.A. (1998). Configuration Prefetch for Single Context Reconfigurable Coprocessors. In: *6th International Symp. on Field Programmable Gate Arrays*. California, pp. 65–74.
12. Hauser, J.R., Wawrzynek, J. (1997). Garp: A MIPS Processor with a Reconfigurable Coprocessor. In: *IEEE Symp. on FPGAs for Custom Computing Machines*. California, pp. 12–21.
13. Heinrich K. (1992). In: *MIPS RISC Architecture*. Prentice Hall.
14. Intel Corporation. Intel PCA Application Processors. http://www.intel.com/design/pca/applicationsprocessors/index.htm
15. Jacob, J.A., Chow, P. (1999). Memory Interfacing and Instruction Specification for Reconfigurable Processors. In: *ACM/SIGDA Seventh International Symposium on Field Programmable Gate Arrays*, Monterey. California, pp. 145–154.
16. May, C., Silha, E., Simpson, R., Warren, H. (1994). In: *The PowerPC Architecture*. Morgan Kaufmann Publishers, Inc.
17. Miyamori, T., Olukotun, K. (1998). A Quantitative Analysis of Reconfigurable Coprocessors for Multimedia Applications. In: Pocek, Kenneth, L., Arnold, Jeffrey, M., eds. *IEEE Symposium on FPGAs for Custom Computing Machines*. California, pp. 2–11.
18. Motorola. Motorola 68000/Coldfire Family. http://e-www.motorola com/webapp/sps/site/homepage.jsp?nodeId = 03M0ylgrpxN
19. Razdan, R. (1994). PRISC: *Programmable Reduced Instruction Set Computers*. PhD thesis. Cambridge, MA, USA: Harvard University.

20. Schmit, H. (1997). Incremental Reconfiguration for Pipelined Applications. In: *IEEE Symposium on FPGAs for Custom Computing Machines*. California, April, pp. 47–55.
21. Seal D. (2000). *ARM Architecture Reference Manual*. Addison-Wesley.
22. Rathnam, S., Slavenburg, G. (1996). An Architectural Overview of the Programmable Multimedia Processor, TM-1. In: *Proceedings of COMPCON '96*, IEEE, pp. 319–326.
23. Trimberger, S.M. (1998). Reprogrammable Instruction Set Accelerator. U.S. Patent No. 5,737,631.
24. Vassiliadis, S., Wong, S., Cotofana, S. The MOLEN ρμ-Coded Processor. In: *Proc. of the 11th Intern. Conf. on Field-Programmable Logic and Applications (FPL2001)*, pp. 275–285.
25. Weaver, D.L., Germond, T., eds. *The SPARC Architecture Manual (v9)*. Prentice Hall.
26. Wilkes, M. V. (1951). The Best Way to Design an Automatic Calculating Machine. In: *Report of the Manchester University Computer Inaugural Conference*, July, pp. 16–18.
27. Wittig, R.D., Chow P. (1996). OneChip: An FPGA Processor with Reconfigurable Logic. In: *IEEE Symposium on FPGAs for Custom Computing Machines*, pp. 126–135.
28. Wong, S. (2002). Microcoded Reconfigurable Embedded Processors. PhD thesis, Delft University of Technology, Delft, The Netherlands, December.
29. Xilinx Corporation. Virtex-II 1.5V FPGA Family: Detailed Functional Description. http://www.xilinx.com/partinfo/databook.html
30. Xilinx Corporation. Xilinx MicroBlaze. http://www.xilinx.com//xlnx/xilprodcatproduct.jsp?title=microblaze

Index